T0350987

Teaching Mathematics at a Technical College

Not much has been written about technical colleges, especially teaching mathematics at one. Much had been written about community college mathematics. This book addresses this disparity.

Mathematics is a beautiful subject worthy to be taught at the technical college level. The author sheds light on technical colleges and their importance in the higher education system. Technical colleges are more affordable for students and provide many career opportunities. These careers are becoming as lucrative as careers requiring a 4-year degree. The interest in technical college education is likely to continue to grow.

Mathematics, like all other classes, is a subject that needs time, energy, and dedication to learn. For an instructor, it takes many years of hard work and dedication just to be able to teach the subject. Students should not be expected to learn the mathematics overnight. As instructors, we need to be open, honest, and put forth our very best to our students so that they can see that they are able to succeed in whatever is placed in front of them. This book hopes to encourage such an effort.

A notable percentage of students who are receiving associate's degrees will go through at least one or more mathematics courses. These students should not be forgotten about—their needs are similar to students who are required to take a mathematics for a degree such as the bachelor's degree.

This book offers insight into teaching mathematics at a technical college. It is also a source for students to turn toward when they are feeling dread about taking a mathematics course. Mathematics instructors want to help students succeed. If they put forth their best effort, and we ours, we can all work as one team to get the student through the course and on to chasing their dreams.

Although this book focuses on teaching mathematics, some chapters expand to focus on teaching in general. The overall hope is the reader will be inspired by the great work that is happening at technical colleges all around the country. Technical college can be, and should be, the backbone of the American working class.

Teaching Mathematics at a Technical College

Zachary Youmans

CRC Press
Taylor & Francis Group
Boca Raton London New York

CRC Press is an imprint of the
Taylor & Francis Group, an **informa** business

A CHAPMAN & HALL BOOK

First edition published 2023
by CRC Press
6000 Broken Sound Parkway NW, Suite 300, Boca Raton, FL 33487–2742

and by CRC Press
4 Park Square, Milton Park, Abingdon, Oxon, OX14 4RN

CRC Press is an imprint of Taylor & Francis Group, LLC

ISBN: 978-1-032-26243-7 (hbk)
ISBN: 978-1-032-26242-0 (pbk)
ISBN: 978-1-003-28731-5 (ebk)

DOI: 10.1201/9781003287315

Typeset in Minion Pro
by Apex CoVantage, LLC

Contents

Preface

MATHEMATICS. JUST THE UTTERANCE of the word frightens people as if it were Halloween night. However, it does not have to be that way, especially at a technical college where students are more than likely non-mathematics majors. The teaching of mathematics at a technical college is an interesting story, one that I hope to lay out in this book.

Not much has been written about technical colleges, especially teaching mathematics at one. Just a quick Google search of "technical college math teaching book" shows many textbooks that can be used in classrooms but none on how to be successful as a technical college math instructor. Replace the words *technical college* with *community college*, however, and many results are returned. What makes a technical college so much different than a community college, you may ask? That question and more will be answered in Chapter 1.

By writing this book, I hope to start shedding light on technical colleges and their importance in the higher education system. Not only are technical colleges affordable for students just leaving high school, but they also provide many career opportunities to students who thought they could never gain a degree higher than a high school diploma or GED. Speaking of a high school diploma, many technical colleges have an adult education division that helps adults gain a GED if they failed to receive their high school diploma. Technical colleges also play a role in the community in which they service. Here at Central Georgia Technical College (CGTC), the Economic Development division partners with businesses across the 11-county area we service, helping them train and hire employees. We also offer dual-enrollment courses to high school students wanting to earn college credit before high school graduation.

Mathematics, and general education altogether, play a major role in the education students receive not only at technical colleges but in all forms of higher education as well. The majority of students who are receiving associate's degrees will go through at least a few general education classes: mathematics, science, English, and humanities. Just to give you some sort of idea about how many students pass through our doors here in the mathematics division at CGTC, in the fall of 2021 there was an unduplicated headcount of 9,613 students taking a course. The number of students taking a mathematics course at CGTC (disregarding some high school dual-enrollment courses, which would make the number go even higher!) was 1,632. That is approximately 20% of the students at CGTC!

So, while students are "frightened" by the word *mathematics*, a lot of them still have to end up taking a mathematics course when they attend a technical college. They are the biggest reasons why I wanted to write this book; they should not be forgotten about. They deserve as much of an education as those who attend community colleges and 4-year universities whether public or private. They deserve the best, not only out of their mathematics instructors but also out of their English instructors, their nursing instructors, their welding instructors, and instructors in so many other disciplines that technical colleges provide.

I certainly try to keep the conversation geared toward mathematics at technical colleges; that is the name of the book, of course! However, I think it is important to realize while teaching mathematics and teaching English are two different topics, the development of teaching in both subjects can be similar. So, in some chapters in this book, I go away from just teaching mathematics at technical colleges and focus on teaching at a technical college in general.

Let me give one such example of this. In Chapter 6, titled "The Word Easy," as a person with a mathematics degree, I perceive finding the square root of 25 to me as easy. So, if I am teaching a Foundations of Mathematics course, which here at CGTC is a beginning-level math course, I might slip in front of my students and say, "Well, finding the square root of 25 is easy! It's 5, of course!" (and trust me, I have done this plenty of times while teaching). Two phrases here are of importance, "easy" and "of course." For students who are learning about the square root for the first time, this may be neither easy nor obvious.

However, this lesson should not be learned just for mathematics instructors. It is a lesson that can be taken by English instructors who are having students pick out nouns from sentences or nursing instructors teaching how to do CPR on an adult or welding instructors showing a student how to be safe while working. These topics and techniques may seem easy and obvious to the instructor, but the instructor has many years of experience and possibly multiple degrees in the subject. For a student, this may be the first time they are seeing these topics in their life. It is not fair to them when we think it is easy, and when we teach it as such, we might lose the student in the woods. Not to get too far ahead of myself, however; more on the word *easy* in Chapter 6.

Overall, though, I hope you, the reader, will be inspired by the great work that is happening at technical colleges all around the country, not just at CGTC. Technical colleges can be, should be, and are the backbone of the American working class, producing thousands of graduates every year who will go work in the health care system, plumbing, electricity, HVAC systems, and multiple other blue-collar jobs. We as instructors have a huge part to play in their success, whether it is being an instructor in their major programs or a general education instructor such as myself in mathematics.

Mathematics is not a subject to be nervous or scared about. Just like all other classes, it is a subject that needs time, energy, and dedication to learn. For an instructor, it takes many years of hard work and dedication just to be able to teach the subject. So as a student, it should not be expected that they will learn the mathematics overnight. As instructors, we need to be open and honest and put forth our very best to our students so that they can see that they are able to succeed in whatever is placed in front of them.

So let this book not only be an insight into what it is like teaching mathematics at a technical college, but a source for students to turn toward when they are feeling dread in taking a mathematics course. It is my hope they read (especially Chapter 7) and be inspired that all we want as mathematics instructors is for them to succeed. If they put forth their best effort, and we ours, we can all work as one team to get the student through the course and onto chasing their dreams.

Thank you to you, the reader, for taking the time to read this book. Keep in mind as you read that mathematics is a beautiful subject worthy of being taught even at the technical college level. Hopefully, by the end, you will be inspired to keep doing great work in whatever you are doing. Whether that be mathematics, teaching mathematics (it doesn't have to be at a technical college!), or a student just curious about what it is like to teach mathematics, I wish you all nothing but the best.

Zachary Youmans

Acknowledgments

T HIS BOOK, AND REALLY a large part of what I have done the past 6 years, would not be possible without my wife, Kylie. We have been through many adventures: from 4 years at the University of Northern Iowa to moving halfway across the country to a place we have never been, Logan, Utah. Now, we get the exciting times of living in Georgia, watching our daughter grow up to be a strong, confident individual. So, to you, thank you for your hard work, your dedication to my career, and being my partner in life.

Another big thank you goes to the publisher of this book, CRC Press/ Taylor & Francis Group, and its senior editor, Bob Ross. As this is my first journey into the world of book writing, he has been nothing but the best in terms of answering questions and navigating the waters of publishing. If you are thinking about writing in the mathematics realm, let me be the first to recommend Mr. Ross and CRC Press. It has been nothing but a pleasure to work with you all, so thank you.

Finally, I want to acknowledge all the individuals who have made an impact in my mathematical and teaching journey. Dr. Michael Prophet, Dr. Marius Somodi, Dr. Ian Anderson, Dr. Zhaohu Nie, and Dr. Latricia Hylton, thank you to all of you for making such an impact on my life. I've learned so much from all five of you, plus all the other professors I have had at both the University of Northern Iowa and Utah State University. Every day, I strive to be as great of an instructor as all of you.

Biography

Zachary Youmans is the current program chair and mathematics instructor at Central Georgia Technical College, a 2-year technical college in Macon, Warner Robins, and Milledgeville, Georgia. As program chair, he oversees 9 full-time faculty members and 12 adjuncts who all serve anywhere between 1200 and 1600 math students in the spring and fall semesters. Before becoming a math instructor, he served as a graduate teaching assistant at Utah State for 2.5 years and a mathematics and statistics tutor for 3 years at the University of Northern Iowa. Alongside his wife, Kylie, and daughter, Aubree, he has two cats, Leo and Miko.

Figures and Tables

CHAPTER 1

An Introduction to a Technical College

The only thing that interferes with my learning is my education.

—ALBERT EINSTEIN

At the ripe old age of 25, I had just gotten done with my master's thesis titled *Some Examples of the Liouville Integrability of the Banded Toda Flow*. My dream: to become a professor at a 4-year university, where I would do mostly teaching with a little research. The research side of mathematics, it was sort of fun and interesting. With the support of my advisor, Dr. Zhaohu Nie, I got through the master's thesis and defended it (virtually, I might add!) successfully in the spring of 2020. In all honesty, teaching was and is still more of my flavor. It's what I wanted to do since high school.

I started at the University of Northern Iowa in Cedar Falls as a math education major wanting to teach high school. Then, after taking a few strictly education courses, I removed the "education" part of the major to become simply a mathematics major with a minor in educational studies. Not that I didn't want to teach mathematics anymore; I was just not very interested in the lesson plan side of teaching, and I preferred working with college students.

So, 4 years of undergraduate work and another 2 years of graduate work at Utah State University led me to receiving both my Bachelor of Arts and

DOI: 10.1201/9781003287315-1

Master of Science in Mathematics. However, as many readers are aware, the spring of 2020 was a difficult one. COVID-19 swept the nation, and it shut down schools and businesses alike. For me personally, I grew tired of the graduate program. Don't get me wrong, I was enjoying my experience at Utah State. However, I wanted to start a family with my wife, be making more than just $15,000 a year as a graduate student, buy a home, and do what I love to do, teach. This means the community college route. While there may be stigmas behind teaching at community colleges rather than universities, I didn't really care.

The job hunt began. During the summer of 2021, I received an interview at a community college in my hometown of Burlington, Iowa. Much to the dismay of my family living back in Iowa, I did not receive a job offer. Oh well, I thought, I will at least start the PhD program at Utah State (which I was accepted into); that way, I will have no regrets in not at least trying to start the program. A couple months go by, and I again start applying to many more jobs around the country. My wife and I decided it did not matter where we relocated to.

I applied to Central Georgia Technical College (CGTC), never really thinking about the technical college route. In Iowa, there are not many technical colleges, nor are there in Utah (eight in total in Utah). However, in Georgia, there are 22 technical colleges all governed by the Technical College System of Georgia, TCSG for short. I slightly knew about technical colleges and their mission, but I was for sure thinking I would be teaching at a community college.

Interviews came around; round one was being questions asked by panel members. Questions were asked such as "Why do you want to work at CGTC?" and "How does your experience align with the mission of the technical college?" Then, after getting through round one, round two was a 20-minute teaching demonstration on the properties of logarithms, which is taught in many of the algebra courses at CGTC. After the demonstration, a 45-minute phone conversation with the vice president for academic affairs was completed. A few weeks later, I got the job as a full-time mathematics instructor at CGTC.

I tell this story not because you really care about my life story. I tell this story because of the great unknown which are technical colleges. I never foresaw myself at a technical college, but here I am teaching and working at a place I love. There are many stories such as mine, where faculty and staff never envisioned working at a technical college but they love every second of it.

Take Carrie Dietrich, for example. Carrie is the director of the Putnam County Center at CGTC. Just a little background on CGTC, we service 11 counties in the middle of Georgia, one of those being Putnam County. Eatonton is the largest town in the county, with a population of 6,480. Ms. Dietrich does a lot of work for the college and wears many hats. Not only does she direct the operations of CGTC out of Putnam County, but she also is a community leader by reaching out to businesses and inviting them in to gain exposure for the college. She also serves on many local boards, including the Putnam Leadership Board, created to revamp the Putnam Leadership Program.

Her start in education wasn't because she chose to work at a technical college. She started her career in the admissions and career services department at Georgia College in Milledgeville. The director of the department at the time was hired as the director of the Putnam Center of CGTC, and he wanted Carrie to work for him. So, she did. He eventually left, and the rest is history. She was promoted to the director position (without the pay and title, mind you) in 2005. In 2007, the title and pay were awarded to her.

Carrie and I both share similar stories; we did not believe that we would end up in technical education. Alas, here each of us is working at CGTC with one goal in mind: the success and job placement of our students.

The most well-known form of higher education would have to be the 4-year university system followed by the 2-year community college. However, technical colleges, vocational colleges, and trade schools all have their place in higher education as well. I, and many others, consider these forms of higher education to be the backbone of the American workforce, training students and adults alike to work in blue-collar careers that sustain the American society. Welding, plumbing, nursing, business and computer technologies, aerospace, and even public safety can all be careers that begin with an education at a technical college.

In the state of Georgia, the TCSG oversees the 22 colleges that provide technical education to the state of Georgia. CGTC is one of those. As stated earlier, CGTC services 11 counties in the state of Georgia, including Bibb and Houston County, where Macon and Warner Robins are, respectively. As of September 29, 2021, there were a total of 92,909 students attending the 22 colleges around Georgia. CGTC happens to be the largest of them at 9,116 students (the fall 2021 semester actually ended up with a little over 9,600 students!).

Students across the technical college environment are often part-time and many of whom have full-time jobs already, children to take care of at home, and have been out of school for quite a while. Full-time equivalency (FTE) is defined to be those students who are enrolled in 15 credit hours or more, which makes them a full-time student. As of November 22, 2021, out of the 9,611 students who were registered for a course in the fall of 2021, 4,571 were considered FTE for a percentage of 47.56%. This suggests over half of the students were part-time.

While technical education is an important part of the higher education system in Georgia, the foundation for technical education did not begin in Georgia. According to the Association for Career and Technical Education, the first manual training school was established in 1879 in St. Louis, Missouri.[1] CGTC was established in 2012 after the merger of Central Georgia Technical College and Middle Georgia Technical College in order to reduce costs. The original CGTC pre-merger was established in 1962, and Middle Georgia Technical College was established in 1973.

The Association for Career and Technical Education also states that between the 1920s and the 1970s (an era it calls "Coming of Age"), "career and technical education expanded to include adult education and retraining citizens to re-enter the workforce." Colleges in the TCSG follow a similar pattern to those that were established during the "Coming of Age" period in that there are three parts to the technical college: academics, adult education, and economic development. More on the three parts of the technical college later on.

Regarding student academics, the mission of a technical college is to produce career-ready individuals that promote the economic development of the area serviced by that college. We want to make sure that students graduate in a timely matter so not only can they produce to the economy of the local community, but more important, they can also enjoy what they do in their career without the cost of attending college.

TSCG colleges also offer associate's degree programs in general studies as a means for students to transfer to 4-year colleges. In many other states, community colleges offer this to students, but there are only technical colleges in the state of Georgia, so they get the excitement of helping students transfer to universities around the country.

One advantage of attending a 2-year public college over universities and private institutes is the cost of tuition. According to educationdata.org, the total cost of tuition for a 4-year in-state university is $38,320. The total cost of tuition at a 2-year in-state college is $6,744.[2] That is a difference of $31,576, which is quite a bit of money! Students can get the general education core out of the way for a lesser price at a 2-year and have the same quality of education.

Students taking classes at CGTC and other technical colleges have three options to choose from: a degree, a diploma, and a certificate. The differences between each come down to how many classes a student will take that result in the time it takes to complete the degree. An associate's degree program typically takes 2 years and requires the most classes due to the general education requirements brought with them. The associate's degree at TSCG colleges has four general education requirements: language arts and communications, social and behavioral sciences, natural sciences and mathematics, and humanities and fine arts. The number of classes students must take in general education depends on the degree. Almost all degrees need a humanities course, an English course, a mathematics course, and a natural science course such as biology. Some of them also require a general education elective, which the student can choose what general education course they want to take.

The diploma is very similar to the degree, but it tends to have fewer general education courses to take than the degree level. Also, there are typically fewer major courses to take. It usually takes a little less than 2 years to complete. The certificate on the other hand typically requires no general education requirements, and only a few courses to be taken. These typically take a year or less, and quite a few of them are rewarded through taking classes in the degree- or diploma-level areas. For example, if a student at CGTC wanted to get a diploma in automotive collision repair, they need three general education core courses: a professional development class, Fundamentals of English, and Foundations of Mathematics. They then take 20 occupation courses related to automotive collision repair. However, as they take these occupational courses, if they take Introduction to Auto Collision Repair, Automobile Component Repair and Replacement, and Fundamentals of Automotive Welding (all part of the courses they must take for the diploma), they will be awarded the Automotive Collision Repair Assistant I certificate as well. This serves as a great way for students to build their resumes while job hunting during or after college.

Technical colleges, as suggested before, do more than contribute to higher education and the local community than producing college-level graduates. Two other sides to the college are economic development and adult education. Let's start with adult education.

The goal of adult education is to provide the skills necessary for adult learners to compete in today's global workforce. According to the Adult

Education portion of the CGTC website, "In the United States, an estimated 30 million people over the age of 16 read no better than the average elementary child."[3] The website goes on to say that "the abilities to read, to write, to do math, to solve problems, and to access and use technology—today's adults will struggle to take part in the world around them and fail to reach their full potential as parents, community members, and employees." So technical colleges help adults gain the experience they need to succeed.

Adult education also offers GEDs to adults who did not finish high school. Just some statistics for you, according to DoSomething.org, 1.2 million students drop out of high school in the United States per year. They also go on to point out that a high school dropout will earn $200,000 less than a high school graduate over their lifetime. This number jumps to almost $1 million less than a college graduate if a student does not finish high school.[4]

So, for adults who want to go back and gain a GED, technical colleges are there to help provide them that education. They offer classes and resources to help when it comes test time, and they help with job placement after graduation. Also, just by getting exposure to the college could help attract these students to gain a college education. This is a mutual benefit for both the college and the student, hopefully at the same institution. The student can earn more money by completing a college degree, while the college has an increase enrollment numbers, which also means an increase in tuition dollars. Overall, adult education is a crucial part of a technical college's mission to enhance the economic development of its local community, and it is a great place for adult learners to turn to if they are seeking a high school or college degree.

Speaking of adult education initiatives, CGTC has begun a fourth wing (somewhat under the adult education program) of the college aptly named the CGTC Academy. Our technical college wanted an accredited program so that students could earn a high school equivalency diploma. So, they became accredited through COGNIA, a nonprofit accreditor for primary and secondary schools throughout the United States to offer classes to adult education students wanting to earn a high school diploma or to those dropping out of the public education system.

The last part of the technical college system is the economic development division. This side of the technical colleges is the one that partners with the local community and businesses to provide training and resources. Companies will often contract out training and job orientations to the technical college in exchange for the college providing the resources and instructors to do such training.

The college also offers space inside its buildings for companies to use. For example, GEICO utilizes space inside one of the buildings on the Macon campus of CGTC. The college also routinely holds job fairs for companies for students to get connected with employers. Representatives from the economic development division here at CGTC are always constantly reaching out to businesses and local leaders, gaining exposure for the college to lead the way in developing the workforce around the middle Georgia area.

Overall, all three parts of the college work in harmony so that the college runs smoothly and efficiently. It is important as an instructor, for example, to learn about the different parts of the college so that we can direct students to the appropriate resources. Also, as a representative of the college, you never know when you might run into a local business leader or an adult looking to go back to college out at a grocery store or anywhere around town! It's important to always be mindful of how you talk and behave out in public so that you can be the best possible representative for the college.

One part of the academic side of the college I want to talk more in depth about is dual enrollment. For those who are unaware, dual enrollment is a program through which high school students can gain college credit by taking college classes while still in high school. The classes are taught either by local high school teachers who have the appropriate certification and are hired on as adjuncts (part-time instructors) at the college or by full-time instructors who work for the college. Instructors will either travel to the high school and use their facilities to teach the class or the high school provides transportation for the students to attend classes on the college campus.

There are multiple advantages for a student taking dual-enrolled classes. First of all, college credit is gained if the student passes the course, which means less time for a student if and when they do attend college (although, on the flip side, if a student does not pass the course their college GPA will be affected in a negative way). This means the cost of tuition would be less on the student since graduating college would be quicker. Also, tuition at the college the dual-enrolled student is taking the course from is free for the student. Here in Georgia, this comes with a limitation of up to 30 credit hours, anything past that must come from out of pocket. Also, Georgia high school students are only allowed to take dual-enrolled courses if they are a junior or above. Some exceptions can be made for sophomores taking a technical course.

For the college, this is advantageous because students have a better probability to attend the same college they received college credit from when they were dual-enrolled. Katherine Kinnick, a professor at Kennesaw State University, states that colleges can use dual enrollment to "increase the diversity of their student bodies." She goes on to say that research within dual enrollment has shown "correlations between program size and perceptions of benefits to the institution," and that "directors of programs with the largest enrollments were more likely to agree that dual enrollment benefits the institution by enhancing student recruitment."[5]

So, dual enrollment is a win–win for both the high school student and the college. It is also a win from a local business side. Businesses are able to recruit students out of high school since the student can gain technical certificates through the college while in high school.

One drawback of dual enrollment on the college side is faculty buy-in. From the same article written by Kinnick, she states that both college administrators and college faculty alike share concerns about the quality of courses taught in high schools by high school faculty.[5] Here at CGTA, many faculty also express concern for students' work ethic. They are attending these courses for free, so in many faculty members' eyes, they do not try as hard as tuition-paying college students.

There are a few ways for technical colleges to help alleviate these concerns. First, at CGTC, when we interview potential high school teachers for teaching dual enrollment, we ask them if they have taught dual enrollment before and ask if they are okay with following a strict structure that is given by the college. State standards dictate what should be taught in the course, and high school dual-enrollment instructors need to teach to those college standards. Also, the college dictates that the same midterm and final be given out in the dual-enrolled classes at the high school that would be given out in a course taught at the college. We also perform a classroom observation within the first semester of a teacher being hired on to make sure they are a good fit for the college.

In terms of student work ethics, there is no evidence to show that high school students do not try as hard as college students. In terms of my own experiences, some of my best students came from the high school ranks. There are many other factors at play for high school students including home life, work in other classes, and jobs. None of these factors necessarily mean that dual-enrolled students will be "bad" students.

Take my own high school learning experience as an example. During my time at high school, I worked at a local grocery store named Hy-Vee in the Chinese section at about 20 hours per work. On top of that, I took four Advanced Placement (AP) courses, AP Chemistry, AP Calculus AB, AP Literature, and AP US History. AP classes are another way a student can gain college credit by taking the course and then taking an AP exam. Most colleges accept AP exam scores of 3 or higher to transfer in for college credit. On top of that, I was taking other classes such as English, speech, social studies, psychology, and even a dual-enrollment humanities course. How did I perform with all these stressors on the AP exams? Well, on the calculus one I received a 4 (go figure!). On the chemistry exam, I received a 3, but Northern Iowa wanted a 4 for credit, so I never received college credit for that course. On the US history and literature exams, I received a 2. I wouldn't call myself a bad student; I worked hard in each class. But with many outside stressors (some of which were personal), I did not perform up to expectation.

Let's now talk about some numbers for dual enrollment. To give you an idea of how big dual enrollment has gotten, take a look at some of the state of Georgia and CGTC's numbers.

In the fall of 2020, CGTC had 2,199 dual-enrolled students compared to 5,941 traditional students. This means that 27% of students were dual-enrolled. This number significantly increased in the fall of 2021. In the fall of 2021, 4,054 students were dual-enrolled whereas the number of traditional students was 5,112. Based on overall enrollment, 42% of students were dual-enrolled. In the state of Georgia overall, 21,711 students were dual-enrolled in the fall of 2021, with a traditional enrollment of 71,198, making it so that 23% of students were dual-enrolled. This means that approximately one-fourth of the total enrollment at technical colleges in the state of Georgia were high school students earning college credit. As one can see, dual enrollment plays a big part in the overall mission of the technical college.

Another aspect of technical colleges that can be seen here in the state of Georgia is their partnerships with the court system. CGTC and the other technical colleges around the state partner with correctional facilities to offer inmates a chance to earn a college degree before they are released from jail. Not only does this help inmates find a job after being released, but we also teach them how to function in today's global environment. This provides a lower chance of the individual returning to jail in the future.

Take, for example, an implementation study done in Lancaster, Pennsylvania. Lancaster–Lebanon Intermediate Unit 13 (IU 13, for short), a regional education agency in Pennsylvania, serves Lancaster County Prison by providing adult education and GED classes to inmates. According to the study, 85% of the prison's population is made up of pretrial detainees and parole violators. Inmates can also take classes at the Pennsylvania CareerLink of Lancaster County, which is a career center for these individuals. In the first year of the study, 120 individuals with criminal histories took classes; 50% of students completed at least one level of basic skills instruction at the prison, and 39% did the same at CareerLink. Among IU 13 GED graduates, 42% entered occupational skills training at CareerLink or enrolled in postsecondary education, 25% of those who served at CareerLink obtained employment while only 11% were reincarcerated.[6]

At CGTC, of the 9,613 students who were taking classes, 434 of those were students from the department of corrections. While that number is definitely lower than those of regular students and dual-enrolled students, individuals who are incarcerated are getting second chances. They deserve as much of an education as anyone, and through skills training, these individuals are rehabilitated back into the workforce.

Technical colleges, while not to the scale of 4-year universities, also offer student clubs and athletics. This could be a disadvantage for students wanting to go to a technical college since many students want to attend and participate in student organizations. Options are limited at technical colleges, but there are a few to choose from. For example, at CGTC, we have Glee Club, REACH (an organization centered on raising the college completion rate for male students of color), and the Student Government Association. In total, there are eight student organizations at CGTC.

Athletics-wise is another disadvantage for students who are looking to attend many sports-related activities compared to a 4-year institution. At CGTC, we have men's and women's basketball and men's and women's cross-country, and that is it. Four-year universities could offer students football, soccer, volleyball, wrestling, and so many other collegiate sports. So, while students can participate in athletics, it is nowhere near the size or scale of university-level athletics.

One other aspect a student may seem disappointed about when attending a technical college is the dorm life. Technical colleges are often commuter schools, meaning there are no dormitories on campus. Everyone lives off-campus. Many students in college look forward to dorm life, where they can socialize, make friends, and plan activities. I remember my own dorm life back at the University of Northern Iowa, where our hall would often attend football games together and play card games in the evenings. This also hurts the attendance rates of students in class, since even a minor rainstorm could prevent students from attending a lecture if they are afraid of driving on rain-soaked roads. However, CGTC is in the process, as of writing this book, of building dorms with the help of outside health organizations.

Now that we have an understanding of what technical colleges are all about, let's now go into the differences between community colleges, technical colleges, vocational schools, and trade schools. Community colleges and technical colleges are interchanged often, but it would be a mistake to call one the other. While both offer 2-year degrees and are usually around the same cost, community colleges often focus on general education. Community colleges offer students classes in general education so that when students earn an associate's degree, they can transfer to a 4-year university with their general education core completed. While technical colleges do the same, technical colleges are usually more focused on career development, which can be seen through the mission statement. Students tend to take fewer general education classes and more career-building courses that teach them how to perform specific jobs.

Trade schools and technical colleges are also often confused. However, according to the Center for Employment Training, trade schools are more hands-on. Technical schools usually have more classroom lectures and simulated job training while trade schools solely focus on "hands-on careers that require a base level certification or a specific number of on-the-job supervised hours to enter."[7] Trade schools are often shorter in length, and they do not usually offer associate's degree programs like technical schools. This is similar to vocational schools; vocational schools also offer certificates to their students and do not usually offer associate's degree programs.

Since this book is about teaching mathematics at a technical college, let's look at what the differences are between mathematics at a 4-year university and mathematics at a technical college. While the standards for each math course between the university and the technical college are of the same caliber, the differences lie in what type of classes are taught. In the state of Georgia, technical colleges have mathematics courses that go through Calculus II, which talks about integrability techniques, sequences and series, and a little of vector calculations. The majority of students (I go over the numbers in a later chapter) at technical colleges, however, take developmental math courses or college algebra.

Four-year universities do teach developmental math courses and college algebra but not to the scale of technical colleges. They also have higher-numbered courses in mathematics, such as Linear Algebra, Calculus III, and junior- and senior-level mathematics courses, such as differential equations and number theory. While these are wonderful classes to take, in my opinion, there would be nowhere near the enrollment to maintain the class at the technical college level. In the spring of 2022, we had only 15 students take a Calculus II course that was being offered. Now, imagine how many of those 15 would venture into the realm of higher-level mathematics. Probably not that many.

The main goal of a mathematics instructor at a technical college is for the student to pass the class so they start taking their major courses and graduate on time. My secondary goal, however, is to show a little bit of why and how mathematics can be a beautiful subject to study. If, at the end of the semester, they walk out of the class learning a thing or two from me on why mathematics can be useful to them, I will be a happy camper.

I want to give some statistics on the successfulness of the technical college, more specifically how successful CGTC is at producing graduates and the local economic impact the college provides. These statistics from CGTC were given in the State of the College address from 2021, which can be found on the college's website and YouTube page.

While CGTC seems to be a small technical college in the middle of nowhere Georgia, the local economic impact the college provides to our 11-county service area is to the tune of $349.6 million. We employ more than 1,000 full-time and part-time staff, and this constitutes around a $46 million payroll. CGTC partnered with 254 companies to serve 15,464 incumbent workers. We delivered 3,858,732 hours of customized training to our business and industry partners. The VECTR Center, a center partnered with the college to provide services to military individuals, provided 20,287 services to 3,514 unique individuals. As a side note, its goal is "to successfully transition veterans and their families into Georgia's public colleges, universities, and the state's workforce."[7] Our adult education side awarded 314 GEDs, with a total number of 1,396 adult education students served. Our graduation rate is 81%.[8] These are absolutely wonderful numbers, and I hope it goes to show while technical colleges may be forgotten about, ours and so many other technical colleges around the country are having a huge impact on local communities.

Overall, a technical college is a great resource for students to earn a 2-year degree so they can pursue their dreams. I get so excited teaching at a technical college because it is easy to see how we make a difference in the lives of our students. When I see my students walk across the stage at graduation, I know that they will go on to do great things in whatever career choice they made. Higher education should be accessible to all, and with the cost-effectiveness and the length of programs they offer, technical colleges help achieves that goal.

REFERENCES

1. *History of CTE.* Association for Career & Technical Education, 7 Dec. 2021, www.acteonline.org/history-of-cte/
2. Hanson, Melanie. (2022, Jan. 27). *Average Cost of College & Tuition.* EducationData.org. https://educationdata.org/average-cost-of-college
3. *About Adult Education.* Central Georgia Technical College. www.centralgatech.edu/adult-education/about-adult-education
4. *11 Facts About High School Dropout Rates.* DoSomething.org. www.dosomething.org/us/facts/11-facts-about-high-school-dropout-rates#fnref1
5. Kinnick, K. N. (2012). The Impact of Dual Enrollment on the Institution. *New Directions for Higher Education*, 2012(158), 39–47. doi:10.1002/he.20013
6. U.S. Department of Education. (2015). *Reentry Education Model Implementation Study: Promoting Reentry Success Through Continuity of Educational Opportunities.* https://www2.ed.gov/about/offices/list/ovae/pi/AdultEd/reentry-education-model-implementation-study.pdf
7. *About GA VECTR.* Georgia VECTR Center. https://gavectr.org/about/
8. *State of the College.* Central Georgia Technical College. www.centralgatech.edu/about-cgtc/marketing-pr/state-of-the-college

CHAPTER 2

Know Your Students

Every student can learn, just not on the same day or in the same way.

—GEORGE EVANS

Every student can learn. I love that part of the chapter-opening quote. Every student *can* learn. Whether they have the motivation to learn is one topic of discussion, but every student who walks into our classroom has the capability of learning, no matter their race, gender, religion, or age. It doesn't matter if they are a part of the LGBTQ+ community or not (although this group is greatly underrepresented in STEM fields across the country). All our students have the ability to learn, and they deserve that opportunity to be in the classroom, learning from qualified individuals giving their 100% best.

However, let's not be naïve. While race, gender, religion, age, or wealth should not factor into whether the student can learn, should not and does not are two different concepts. I *should not* be drinking soda, but I do anyway because it provides energy in the form of caffeine. A student *should* still be able to learn even if they have $50 in the bank account, although in reality, that is not the case. They are more worried about how they will feed their families rather than a mathematics assignment worth 0.50% of the grade (that is the percentage each individual homework assignment is weighted in my College Algebra class). As many are aware, however, missing multiple homework assignments results in a poor performance on the assessment that is given at the end of the chapter, and that results in a failing grade for the student. Can you blame your student for that failing grade, though?

DOI: 10.1201/9781003287315-2

I want to provide this chapter in this book to give a glimpse into the type of students that are attending technical colleges. Knowing the type of students that are walking through our doors provides a huge opportunity to try and reach every last student, no matter how hard their situation may be. Knowing our students provides an inclusive classroom, where we can teach o our students in the most proficient way possible. Let's take a look to see different ways we can get to know our students, and what difference it could make in the classroom.

1. Talk with your students before or after class about what is going on in their lives.

The best way to get to know your students is to actually talk to them about what is going on in their lives. While this may seem obvious, I have met and have taken classes from some professors who *never* talked with their students except when they would teach us. For example, I took a class from a professor (he will remain anonymous) in my undergraduate degree mathematics course titled Calculus II.

This professor walked into class (exactly on time, I mind you, which was quite impressive), picked up the chalk, and lectured to the chalkboard for 50 minutes until it was time to leave. I say he lectured to the chalkboard because he *never* turned around to even acknowledge students. If you had a question, oftentimes this professor would get frustrated thinking that the question would have already been answered by him during the lecture, or in a previous math class. You would often hear the professor say, "You should have learned that before!" and just continue on with the lecture without answering the question.

If you don't believe me, go look at this professor's Rate My Professor score! I know, this wouldn't be a good book if Rate My Professor wasn't mentioned at least once. A lot of times, websites like this are used by students to let out their frustration on the professor, so oftentimes, you see bad scores. However, I do use it for myself, especially if there is useful advice that I can take from it to further better my teaching. For this professor in particular, comments are very similar to my own experience. Comments about being quiet, always talking to the chalkboard, and just overall being a bad professor.

Now, I don't like talking about any professor and how they teach. Each instructor has their own unique way of teaching, and that should be celebrated. However, if the majority of students are giving constructive criticism about your teaching that you can build and improve upon, an instructor needs to be adaptive to those changes.

Not only that, an instructor represents the department and the college while teaching in front of a group of students. So, if students are constantly giving negative feedback to the instructor and the instructor does not change anything about the way they teach, students may be reluctant to take classes from the department. Yes, at technical colleges, most students are taking mathematics classes as a general education requirement, but I would much rather have a group of students who, on the first day of class, are excited about learning from me rather than dreading the semester.

Also, students do look at websites such as Rate My Professor to pick out their professors (something I even did during my undergraduate work!). So, if an instructor represents the department and the college, and that instructor is constantly receiving negative feedback each and every semester, students might broaden their opinions to the entire department, not just one instructor. If one instructor is bad, that means they all must be bad, right? Of course not. However, students may feel that the department as a whole is lacking when the students feel the department is doing nothing regarding their feedback. This, in turn, means students will stop or give false feedback due to a lack of trust. Personally speaking, as an instructor I am always wanting new techniques and new ways to teach my students. I listen to their feedback, and I try to implement things students are wanting in their classrooms. My goal is to create a positive learning environment for my students, and listening to them is one way to achieve that goal.

As a student during my undergraduate degree, I was always looking to connect with faculty members I took classes from as well. Students at technical colleges do the same. Just by going in and talking with your students before or after class you can get to know them and their learning habits. This helps develop a rapport with them, and sometimes students will, later on in their journey, come back to you seeking recommendations.

For example, I had a student in a College Algebra course who was a fantastic leader in the classroom. I often went to class a little bit early and talked with her about what her plans were for her future. She was a dual-enrolled student, so she was looking to attend a 4-year after graduating high school, knowing that many of her dual-enrolled college credit would transfer. After the course was over, she asked if I could provide a letter of recommendation for her when she submitted applications to both colleges and scholarships. Without taking the time to get to know this student, it would have been difficult to provide such a recommendation. However, I was happy to provide this recommendation to her, and it makes me feel pretty good knowing that when she does get accepted into a college, I had a little part in that decision.

One last implication of talking with students before or after class is the fact that socializing is a great way to keep enjoying the job even after many years. In Chapter 7, I give the numbers on teacher burnout rates (which are quite high), but for now, it's important to realize that instructors do get burned out for a number of different reasons (reasons are also given in Chapter 7). At least for me, it excites me when students walk into class asking me about how my day is going. It excites me when I, in turn, ask them how they are doing, and they are honest with me because I have taken the time to develop a relationship with them. This honesty may come in the form of struggling with a concept being taught about, something that is happening at home, or maybe they are doing just fine and do not have any other comments to make.

2. Knowing our students makes instructors seem more human.

This seems like a weird statement to make; of course, me as an instructor is human! What makes us a human? I can taste, hear, smell, feel, and see. I have some sort of consciousness. Although I am not a psychologist, I believe I am human! So, why would students feel as if I am not one? First, I am in a position of power. I have the power to fail the student, pass the student, and, as long as I am following state, college, and department standards, teach however I want. I have the power of knowledge. I have a master's degree in a subject that people are terrified of, and most cannot comprehend how one could even like mathematics enough to get a degree in it. So, I get looked at as either someone above the rest or someone who is crazy enough to teach a subject that is ostracized.

Honestly, I would rather have neither. I am just a normal guy living a semi-normal life. I have a house, a car, a wife, a daughter; I work in my dream field; and I get to work with students from different backgrounds. However, students do not know this unless we tell them! I don't mind sharing details about my life to my students, so they see me as more human.

For example, the fall of 2021 was a difficult one for me. Not only was my wife and I raising our first child, but my wife was diagnosed with postpartum depression and anxiety, which caused her to lose her job and cut our income in half. This was, and still is, affecting us to this day as of this writing. I told this to my students at about the halfway part of the semester when a lot of students were struggling with not just coursework but life events as well. My goal was for students to see that while life was throwing me difficult challenges, I still showed up to teach them all with a smile on my face, willing and eager to show them the wonders of mathematics. I overcame adversity and preserved through rough times. When our students see that, not only do they view me as more of a human, but they are motivated to keep pushing through as well.

3. Some students are in difficult situations, and we must be aware of that.

When we get to know our students, they will in turn start to trust us as the instructor, and they made start to open up about what is going on outside of the classroom. My advice: listen. Just as in the previous point about myself being in a difficult situation makes me more human, students themselves are going through a lot. When we listen to our students, they will trust us and will keep fighting for success. Also, knowing that they are in difficult situations leads me to my next point in that . . .

4. We can provide the correct resources to our students if we know what they are going through.

It's amazing the number of resources that are available to students, and often enough, students have no idea about the resources that are available to them free of use just for being students at the college. A college will have resources not only for academic help but also for other scenarios, such as mental health, veteran services, and career services.

Beginning with academic concerns, I am personally a huge proponent of tutoring. In fact, I was a mathematics and statistics tutor at what was then called the Academic Learning Center at the University of Northern Iowa for 3 years before moving on to graduate school. At Central Georgia Technical College, we have a wonderful Academic Success Center on all three of our campuses: Milledgeville, Warner Robins, and Macon. They also offer online tutoring for students who are off campus.

Unfortunately, there is a bigger-than-expected bias against tutoring. In my experience, students look down on tutoring, thinking they are not smart because they need additional help. This is absolutely not the case. Everyone needs help at some point in their life, whether it is to ride a bicycle or you have questions pertaining to your job. I sometimes need to seek additional help at times from my boss or other colleagues if I am unsure of what to do. There is nothing wrong with this! I would much rather have someone ask a question than to guess and be wrong. Tutoring is similar; it is not the fact you are not smart. It's the fact that the tutor has more experience and wants to help answer your questions. My boss has more experience than me in a lot of different areas, so it is a good idea to ask him questions if I do not know the answer. And that's okay. One of the only ways to learn is to ask questions and to seek guidance when lost.

Students on campuses have a wide array of other scenarios besides academics that might affect them as well, almost too many to count. As a shout-out to my brother-in-law Dr. J. Cody Nielsen, as of this writing the director of the Center for Spirituality and Social Justice at Dickinson College in Carlisle, Pennsylvania. He is a huge proponent of the equity of religious minorities and nonreligious identities on campus. His work with Convergent Strategies is laying the groundwork for universities to start recognizing the need to uphold the rights of religious and nonreligious people on campus. However, this equity should not stop just at universities, but it should spread to 2-year colleges, such as technical colleges, as well. So, resources for students to turn toward if they are feeling ostracized due to their religion and/or sexuality is a must for colleges.

While colleges such as Dickinson have the Center for Spirituality and Social Justice for students to turn to, 2-year colleges probably do not have such resources. This is where we as instructors need to stand up and show our students that we want to be fair and equal in higher education. If a test falls on a date that a student is observing a religious holiday? Move the test for them! If they need to miss class to pray? Make sure they gather the information they need after class! Whatever it takes, instructors need to be cognizant of students' backgrounds in our classes and know of the appropriate resources they can turn to if a student is feeling left out due to their background.

Students also may have a mental health crisis mid-semester. If a student comes forward to you about this, first off, please acknowledge their crisis and thank them for coming forward. I tell students who come to me that I am glad they focused on their mental health first, because without being healthy physically or mentally, one is at a huge disadvantage for being successful. Point them toward counseling services; all colleges should have some form of counseling. At Central Georgia Technical College (CGTC), we offer free counseling services to all our students. Making our students aware of these services shows that we care about them as individuals, and even if a student does decide college is not for them due to a multitude of different reasons, students can feel safe knowing that these resources are available to them.

Another resource that is offered to our students is career services. Career services helps current and graduated students in job searches as well as resume assistance and career counseling. So, if I know a student in my class is about to graduate, I could refer them to career services to get the student to start thinking about resumes and how to apply to careers in the student's chosen career field.

One last resource I want to talk about is disabilities services, or what at CGTC are called special populations. Students with disabilities can and should have equal access to education, and by going through special populations, they can give the instructor special accommodations throughout the semester to help the student. A student does not have to reveal their disability to their instructor per the Americans with Disabilities Act, but honestly, that should not matter. If a student wants an accommodation, I will, 99% of the time, grant it unless it is unreasonable (personally speaking, I have not had any unreasonable requests). I had a student with a disability once tell me that her teachers in high school were not accommodating to her disability. However, we here at the college made every effort to accommodate and now she is applying to 4-year universities with recommendations from myself and other instructors around the college. Stories like this of preservice even with a disability only prove to show that nothing will stop someone from achieving their dreams.

Overall, all colleges are different in terms of the resources they offer to students. Please, make sure you as the instructor are aware of all the different resources on campus so that we as a college can better serve our students.

5. Collecting data at the end of the semester on grades will provide insight into when students are struggling.

We can get to know how our students learn through analytics. If I know the quiz that student's struggle on the most, I can spend more time on that part of the class than others that students succeeded in. Also, a dip in student performance could also mean something happened during that semester. For example, in the spring of 2020, COVID-19 shut many schools down, and schooling went online, so student scores could have changed after that switch from face-to-face to fully online.

Let's take a look at an example of this from two semesters of my own teaching: the spring of 2021 with the fall of 2021. The following data was collected from my MATH 1111: College Algebra course. For perspective, in the spring of 2021, I had 67 students take College Algebra from me (fully online) and 65 students in the fall of 2021 (both online and face-to-face). Both semesters, students went through the same curriculum and standards with seven quizzes. The first chart (Figure 2.1) is for the spring 2021 semester. One note: one of the lines will have one fewer quiz than the others due to the fact we did not reach quiz 7 because of time constraints.

The next chart (Figure 2.2) is for the fall 2021 semester.

These graphs are quite amazing, in my opinion, since each section for each semester follows a similar pattern. Notice that quiz averages were a little lower to begin the semester; then scores went higher for the middle parts of the course and then lowered at the end of the class. There could be a few reasons for this. One, students at the beginning of the semester are still in the "feeling out" period. Especially in the fall semester when there are many new freshmen on campus, they are not accustomed to college life, and this feeling-out period is reflected in their overall quiz scores.

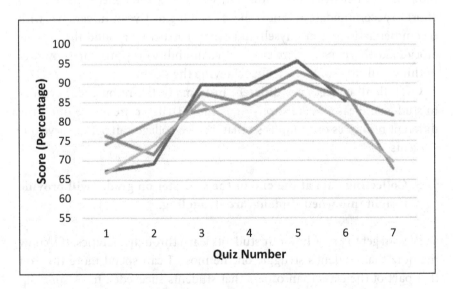

FIGURE 2.1   Quiz scores per section for College Algebra, spring 2021.

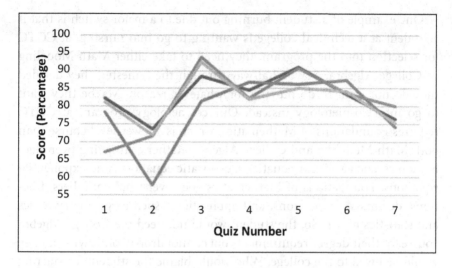

FIGURE 2.2   Quiz scores per section for College Algebra, fall 2021.

As they grow accustomed to college life, how the class is running, and the teacher's grading and teaching style, grades tend to go up until the end of the semester when burnout starts to take place, especially in the spring semester.

Burnout and my earlier point on knowing the difficult situations our students are going through could go hand in hand. I say could because sometimes people just get tired due to how many classes they are taking, how much work is involved in the classes, or maybe they are starting to realize that their chosen career field is not exactly what they want to do.

Speaking of choosing a career field, it is important to realize students might fall out of favor with what they are majoring in at the technical college. According to a report by the U.S. Department of Education, 30% of students earning associate or bachelor's degrees will change their major within the first 3 years of enrollment. More specifically for associate's degrees, that number is 28%.[1] The same report goes on to say that about 10% of associate's degree students change their majors more than once. Just to point out this statistic because I find it interesting; while there are no mathematics majors at the associate's level, at the university level in the year 2011–2012, about 52% of students who were declared a mathematics major switched within 3 years. What does that say about how instructors are teaching mathematics?

One example of a student burning out due to a major switch is that if a student at a technical college is wanting to go into nursing, at CGTC for selection into the program, they need to take either Math Modeling or College Algebra. Maybe halfway through the semester, they decided they do not want to do nursing for whatever reason. Maybe they want to go into Cosmetology instead. Our cosmetology program at CGTC requires Foundations of Mathematics, which is a lower-level course than both Math Modeling and College Algebra. Rather than going over algebra topics, such as linear equations, quadratic equations, and exponential equations, Foundations of Mathematics goes over whole numbers, fractions, decimals, proportions, and applications, such as geometry topics and statistics topics. So, the student would not need the College Algebra course for their degree requirements but cannot drop it; otherwise, money would be owed to the college. Who would blame the student for starting to do poorly in the course? It isn't a requirement anymore, so the student is basically taking it for no reason other than money.

However, burnout can happen for other reasons as well. I would like to give an example of my own story of burnout. I've already discussed some of my story back in Chapter 1, but I would like to expand on it now. During the spring of 2020, I was in my last semester of my master's degree program. One thing about graduate school that is so important is a support system, and that support system for me included another graduate student who was taking the same classes as me, and he was about to write his thesis and graduate as well. Well, mid-February he decided to drop out of school. For me, that was devasting because I lost not only my study buddy but someone I could talk to about what was going on around the department as well. Sure, I still could and still do talk to him, but it is different when he himself wasn't in the program anymore. Then, March hit and COVID-19 forced everyone to go online. I was inside all the time, and I did not have many interactions with any other graduate students. So I became burned out. I struggled with finishing the thesis (which I did successfully defend that same semester). I struggled in finishing classwork (although thankfully the professors were understanding and caring). Thankfully, I persevered and finished the program so that I can be writing this book and teaching mathematics at CGTC.

This perseverance, unfortunately, cannot be said for many other students. The pandemic created a situation where enrollment rates and retention rates dipped. A report by the College Board titled "College Enrollment and Retention in the Era of Covid" states that "student enrollment rates declined more substantially at two-year colleges than four-year colleges." That number, they go on to say, is a decrease of nearly 12%. Retention rates at 2-year colleges the report goes on to list decreased by 4.9%.[2]

Here is another statistic the same report goes on to list that I think is important to realize the barriers students have when attending college. During the pandemic, the rate of enrollment at 2-year colleges declined 16.1% among students whose parents had a high school diploma or less and 12.7% among whose parents had some college experience. They also studied the impact of enrollment regarding unemployment rates and found that at 2-year colleges, enrollment rates declined by 15.1% among students in counties with an unemployment rate of 10.7%.[2]

These numbers are telling because of the disparity in who has access to education. There are many barriers that our students face, and oftentimes, they lead to burnout or just plain dropping out of college. We need to know our students so that we can provide them the resources (as mentioned previously) and the time to help retain them and get them to graduate college.

A significant barrier, and one that we as instructors need to be aware of, is that of our students who live in poverty. According to talkporverty.org, in the state of Georgia, 13.3% of those live in poverty (which by this website is defined to be making $25,925 or less for a family of four). This ranks 37th in the country. In terms of children, 18.5% of children younger than age of 18 live in poverty. Of those who live in poverty, 18.8% are African American, and 9% are white.[3] For students who live in poverty, this a great barrier to their success. They may be working multiple jobs and taking care of children, all while trying to earn a college education that would provide them a better future. And, as I asked earlier, who could blame them for not succeeding in this type of environment? The deck is already stacked against these students. Now, on top of having to figure out how to put food on the table, they also need to figure out tuition, books, and other associated costs that there are in going to college.

There is a fantastic book that is written by Dr. Ruby Payne titled *A Framework for Understanding Poverty: A Cognitive Approach for Educators, Policymakers, Employers, and Service Providers* that I would recommend reading to anyone who is wanting to understand poverty from the side of education. While I could quote many ideas from of her book, there is one line that backs up my point on getting to know your students creates a huge advantage in creating a positive experience for students, and that is on page 87 when she writes, "Teachers and administrators are much more important as role models than is generally recognized." She goes on to say that "the development of emotional resources is crucial to student success" and that the "greatest free resource available to schools is the role modeling provided by teachers, administrators, and staff."[4]

If we do not get to know our students, how are we supposed to be effective role models for our students? It is my opinion that 80% of the battle in a mathematics classroom is for students to buy in to what is being taught, and alongside that thought is the fact that we as instructors need to be role models for our students. Eighty percent is not a factual number, but I say that because a lot of students walk into mathematics classrooms with a negative attitude toward mathematics. How often does one hear "I hate math" or "I am not good at math"? Even throughout the semester, students struggle, not just with the math itself but also with building the confidence to even begin working. An effective role model would be one to where it is okay to make mistakes, and when we face obstacles in and outside of the classroom, we ask questions, plan a way out of the obstacle, and execute the plan. I talk later, in Chapter 7, about how making mistakes can help encourage growth in our students.

Having this effective role model provides the student with a sense of hope and the will to succeed in our mathematics courses. Once we get over that barrier, once we throw out all the myths about mathematics, and once we throw out all the negative attitudes about mathematics, we can forge ahead and actually start to learn the material. Take a student I had for an example; we will name this student Racheal (not her real name). Racheal was down on herself all semester; after every quiz, she would walk into the classroom sad because in her mind, she had failed the quiz. What really happened, however, was that oftentimes, she would be certain that her score was in the 60s or 70s on the quiz, but after I graded the quiz, she would score in the 80s or 90s. I distinctly remember for one quiz, when I told her that she got somewhere in the 80s and not in the 60s like she originally thought, she almost broke down crying right then and there.

With Racheal, I often encouraged her to keep going. She was doing great in the class, so even one little trip-up doesn't mean it is the end of the world. However, her issue wasn't doing the mathematics. I could clearly see she was trying her best to understand. No, it was, in fact, her confidence was getting in the way of her success. Luckily, she was successful in the course, but if I did not take the time to get to know Racheal and her circumstances, she may have decided to stop trying midway through the semester.

When we get to know our students, the world opens up to us as instructors. When we get to know our classroom and those who are in it, we as instructors can walk out of the classroom at the end of the semester knowing that we made a difference in the majority of our students' lives. That is what makes this job worth it in the end for me. I can teach the same class 100 times in my lifetime, but each time I teach it, the students and their circumstances change. So, I am proud to stand up in front of my class and get to know them and teach them a little bit about mathematics. Hopefully, that feeling is the same for you, too, no matter the subject area you are teaching.

REFERENCES

1. U.S. Department of Education. (2017). *Beginning College Students Who Change Their Majors Within 3 Years of Enrollment.* https://nces.ed.gov/pubs2018/2018434.pdf
2. Howell, J., Hurwitz, M., Ma, J., Pender, M., Perfetto, G., Wyatt, J. and Young, L. (2021). *College Enrollment and Retention in the Era of Covid.* College Board. https://research.collegeboard.org/pdf/enrollment-retention-covid2020.pdf
3. *Poverty Rate in Georgia 2020.* TalkPoverty.org. https://talkpoverty.org/state-year-report/georgia-2020-report/
4. Payne, R. (Ph.D.). (2019). *A Framework for Understanding Poverty, A Cognitive Approach for Educators, Policymakers, Employers, and Service Providers.* Aha! Process, Inc.

CHAPTER 3

The University Effect

Every expert was once a beginner.

—RUTHERFORD B. HAYES

The 19th president of the United States is correct; every expert was once a beginner. As brilliant as Albert Einstein and Stephen Hawking were, even they had to start somewhere. This is one of the hardest things to remember for me while teaching a mathematics course, because as much as I love my students, I also love mathematics. It is beautiful. It is elegant. It is full of mystery and suspense. You start to learn this while taking upper-level mathematics courses at the university. I remember one class I took just because it had a cool name to it: Dynamical Systems: Chaos Theory and Fractals. Chaos Theory? That sounds like fun! In reality, it was the fractals that are quite beautiful. Google fractals, and you will see what I mean.

So, while I think certain topics in mathematics are cool, students at technical colleges, candidly speaking, do not. And why should they? Most of them are not interested in majoring in mathematics; they need to take the course for their general education requirement and then move on to their major courses such as nursing, business, or really whatever else the college offers. At Central Georgia Technical College (CGTC), we offer five different fields of study: aerospace, trade, and industry, business and computer technologies, health sciences, public safety and professional services, and general studies. While all of them use mathematics, they do not use mathematics in the way a person majoring in mathematics would use it.

DOI: 10.1201/9781003287315-3

Also, as a setup for the chapter, the difference in student populations between the university and a 2-year college such as a technical college is enormous. Now, I don't want this to be a slam against people who go to 4-year or 2-year schools. Both are wonderful options, but the type of student makes a difference in how a math instructor should teach the class. This fact also contributed to the hard transition between university teaching and technical college teaching.

This creates what I call the university effect: where instructors often get their start in teaching at the university level as graduate students but then transition to a role in which developmental mathematics is favored. Many instructors may have also gotten their start at the high school level, teaching high school–level students until getting a shot at teaching higher education. So, while they may not have gotten the full glimpse of university teaching life, they still received a graduate degree from a university. A transition into higher education will create a sense of disdain toward students at 2-year colleges, many of whom "do not know how to college" as I have heard it described. These students turn assignments in late, wait until the last possible moment to turn in work, and do not study, and many instructors would believe just flat out do not care about their success at the college. I mean, how could they when a lot of students have a GPA lower than 2.0?

This type of mindset is one I want to break instructors of, especially those who are new to technical education or really 2-year education in general. Let me begin in the world with which I am all too familiar: the transition from university-level teaching to technical-level teaching.

Let me take you into the world of Utah State University (USU), a public university with about 27,000 that actually just reached Carnegie R1 classification as of December 2021. Just some background on the university that will be used later: the average age of an undergraduate student is 22,[1] with about 83% of those undergraduates being white.[2] The class sizes I taught in the classroom ranged between 25 and 40.

My very first semester as graduate student, I was what USU calls a recitation instructor for Calculus III. A recitation instructor is someone who helps the main instructor teach the course. Students were required to sign up for recitation sections, and they were essentially study halls for students. Main instructors could give graded quizzes or homework in the recitation as well if they so desired. This gave me my first glimpse into teaching, while I wasn't the main instructor, I still had to guide the classroom through activities designed for Calculus III students.

The following semester I received my first gig as a main lecturer teaching trigonometry. What a moment this was for me! Trigonometry is a class at USU that serves the purpose of precalculus, so the majority of students would go on to take Calculus I. This right here is one huge difference between technical college mathematics and university-level mathematics, but more on that later.

I was then given my own College Algebra course for two semesters, one during the summer and one during the fall semester. Just in this lies many differences between technical colleges and universities: my summer course was a telepresence course, where I broadcasted my course to five other USU locations around the state. During the fall semester, many students were freshmen. Again, more on how this differs from technical colleges later.

I then went on to teach Calculus I for two semesters and Linear Algebra (one of my favorite courses to teach due to how wonderful and useful linear algebra is) for one semester. Linear algebra is not taught at many technical colleges (in fact, none in the state of Georgia). Calculus II is the highest mathematics course that is taught at technical colleges in the state of Georgia, and even that class has low enrollment.

Now for some data regarding CGTC, approximately 47% of students are African American while 44% are white.[3] While there is not a specific number on the average age of an undergraduate, I can reasonably guess it is more than USU's average age of 22. Research is not a part of the job description of most professors at technical colleges, so instructors mainly focus on teaching and service to the college. Instructors at universities may have two or three classes they teach per semester; instructors at 2-year colleges may have five or six classes they teach per semester (except summer).

While these numbers I am giving are specifically for CGTC and USU, I think it is safe to generalize these to universities and 2-year colleges. Generally speaking, students tend to be older at 2-year colleges due to the affordability and accessibility of going back to school. Also, programs are generally much shorter in length, so adults have an easier time of graduating faster than at a 4-year university.

Anyway, this chapter is more about the instructor and not the student side of things so let's get back to that. First, let's talk about the type of students at CGTC compared to universities. Specifically, in mathematics, students at technical colleges will generally not take a further math course unless specified by their major. Oftentimes, they take a developmental math course before even getting into College Algebra. Studies have shown that approximately 59% of incoming students 2-year colleges take a developmental math course at a 2-year college, but only 50% of those students actually complete that developmental math course.[4] At a university, that may be the case with those students taking college algebra in that they are trying to satisfy a core requirement to graduate. However, when students are paying much more money to be sitting in a 4-year university College Algebra course compared to a 2-year technical college College Algebra course, the motivation to succeed may be better.

Trigonometry used to be a class we taught at CGTC, but due to low enrollment in the course, it has not been offered in many years. At CGTC, there were 1,086 students who were enrolled in College Algebra in the fall of 2021 compared to 14 that were enrolled in Calculus I. Just one of my Calculus I sections at Utah State had 39 students. So, one class at USU outnumbered the entire Calculus I population at CGTC.

Generally speaking, students at CGTC take College Algebra (or a few other options such as Math Modeling that are given for more specific degrees) and stop. This was a major difference for me when transitioning from university to technical education is that it is okay for students to stop at College Algebra because they are trying to achieve their dreams. This dream does not involve Calculus I, unfortunately, but not really unfortunately because the world needs nurses, car mechanics, plumbers, and others. And let's be honest: these jobs do not need to know how to take an integral of a specific function, although in my opinion finding the value of an integral is a cool computation to do. They do not need to know what a basis is or a vector space or a differential equation. I loved teaching about what a basis is and using matrices to help find the linear independence between vectors, but that skill is no use to those in technical education. And that is okay. As much as I may want to teach that subject again in my life, and maybe I will get to, when getting into technical education one needs to realize that you are there to help the students succeed. This success is not to get them really good at math, but to help them pass so they can graduate as soon as possible to contribute to the local economy.

So, if you have a degree in mathematics and you really want to teach the upper-level mathematics courses, technical education is not for you. However, I think you would be missing out on some of the best students you could ask for. It is rewarding to watch students who struggled with mathematics their whole life and finally pass the course to start realizing their dream. To watch them walk across the graduation stage, you know you had a hand in that, and yes, it is rewarding for 4-year professors as well, but there is just something special about watching a 50-year-old adult who wanted to make a better life for themselves and their family walk across the stage, degree in hand.

Along that same note, if you are someone who loves mathematics research, technical education is probably not the best place for you. While most everyone does have a master's degree in mathematics at 2-year colleges, you won't be finding much math research going on, if at all. Instructor jobs at 2-year colleges usually focus on teaching, not research. So, if you are in that world, you could do research on your own, but it would be difficult to do collaborations among colleagues. Instead, math education research could be a possibility at 2-year colleges. There is not much research done in mathematics or really any subject at technical colleges or commuter colleges, so if you are interested in that side of research, the 2-year college system might be the right fit for you!

As mentioned earlier, I taught a College Algebra course through a tele-presence system in the summer of 2019. To be more specific on how this worked, USU had cameras and a TV mounted to the back of the classroom where I could see students from across the state of Utah, and they, in turn, could also see and hear me. They also had microphones in case of a question. USU had computers that I could write on the screen with, which became super handy when I was lecturing. USU also hired students to be tech coordinators who ran the technology in the classroom for me, so I never had to worry about it.

Here lies a difference between universities and 2-year colleges: access to resources and money. At CGTC, we also have telepresence classes where I can transmit my in-person classes to others in different parts of the state. However, I don't have a fancy setup; it is a TV, a camera, and my own personal laptop's microphone and writing screen. Instructors can also use the document camera if they do not have the technology. While this is no fault of CGTC since it is a 2-year college with nowhere near the endowment of USU (which stands at about $430 million), the lack of resources is something a new instructor must get used to.

Before I go any further, I do want to say that while it seems like what I have been saying so far is negative, in fact, it is definitely is not. Many smaller colleges are like family; everyone is out to help each other out. So, if technology is a hinderance, someone will be there to give ideas on how to proceed. Or, if a faculty member is struggling in their job and wants to move, CGTC and many other technical colleges do a good job of moving the person to a position that would be the best fit for them. It's all about what environment a certain individual like to be in: whether that is a big-name university with lots of money and resources with strong research ties or a smaller 2-year technical college where teaching and the graduation of students are the focus.

To back up the differences even further, let's take a look at the mission statements of USU and CGTC, since I am using these two colleges as examples. According to the USU's website, the mission of the university is to be "one of the nation's premier student-centered land-grand and space-grant universities by fostering the principle that academics come first, by cultivating diversity of thought and culture, and by serving the public through learning, discovery, and engagement."[5] On the other hand, CGTC "offers credit instruction, adult education, and customized business and industry training through traditional and distance education delivery designed to promote community and workforce development."[6]

CGTC, and really all technical colleges, are designed to help promote the local economy by graduating students at a fast pace and getting them placed in their dream jobs. Alongside this, we focus on adult education and workforce development, while the university is more focused on academics and research. I certainly have already talked a lot about the certain aspects of the technical college in Chapter 1, so I won't dive into too much more about this, but I hope it is clear that if you are an instructor going from the university to a technical college, or vice versa, you would be almost living in two different worlds with two different missions.

I do have to admit though that as someone who did graduate work at an R1 university and taught there as well, it would be hard for me to go back to that setting after having taught at a place like CGTC. As I've mentioned before and I will mention probably a lot during this book, the feeling you get helping students of all ages and backgrounds succeed and watching them graduate is a wonderful one. Everyone at the college wants the students to succeed, and when that happens, we all gleam with pride.

Going back to teaching upper-level mathematics for a second though, if there is something I do miss about the university setting is the higher-level mathematics that were discussed among peers. I still keep in contact with my friends and fellow graduate students back at USU, but it isn't the same being hundreds upon thousands of miles away. Fellow instructors will talk about higher-level mathematics with you, but when everyone is focused on teaching College Algebra and developmental math courses, there isn't much time left over for these higher-level discussions. A friend of mine and I are trying to put together a monthly conference among former graduate students and friends at USU that will help generate discussions that we used to have in the past, and if you are feeling the same way as I do, that may be a great way for you to reconnect as well. It never hurts to once or twice a month to sit down with peers and have discussions, mathematics or not.

I also previously mentioned a difference between teaching at a university and teaching at a technical college is the idea of the freshman. Freshmen at universities tend to be young, fresh-out-of-high-school students who are probably way over their heads being at such a large university. A small fish in a large pond, as they say. Large class sizes, a lot of competition, but also a lot of opportunities for these students exist in the forms of clubs, dorms, sports, and academics. As an instructor, you realize going into freshman-level classes that they are probably not prepared for a college-level class. The rigor it takes to study and prepare for exams is immense, and when freshmen are taking five classes at a time, time can get away very quickly. So, as an instructor, you would need to accommodate your teaching so that these students can succeed.

At a technical college, however, the idea of a freshmen changes. Students at technical colleges come from a wide arrange of generations, so the spectrum of "freshman" is varying. For example, many students come back to college after many years of being out of school. While they have been to school before, times have changed since they last stepped foot in the classroom. For instance, technology in the classroom has changed immensely, especially in mathematics courses. As a way to present an example of this, as previously mentioned, CGTC math courses use Pearson MyMath Lab as its way to deliver homework, quizzes, and exams. I had a student who was in her 50s returning to college and in the previous semester dropped the College Algebra course after a few weeks partly due to technology difficulties. In the next semester when she took my class, I helped her through her difficulties with the technology, and she passed my course with an A. The fun part of that was she still bought a physical copy of the textbook, although MyMath Lab provides an e-text for students to use. Personally, I had no issues with her using the physical copy; in fact, I utilized it at times when students wanted more examples than what I had prepared!

Now, I don't want to assume that all students who are older come back to college having difficulties with technology. That is certainly not true, and I have students older than me who are way better at technology than me, in my late 20s at the time of this writing!

Alongside technology, many students at technical colleges hold full-time jobs outside of the classroom. So, instructors need to be aware that not only are students juggling coursework but also work, possibly children, and other life events that pop up throughout the semester. If a student has not attended college or has been out of the game for a while, it may take a while to get a schedule that works for that student. This may mean late assignments, failing homework grades, or skipping class, but in the end, our goal at technical colleges is to help the student succeed and graduate. Our department does have a late penalty for homework assignments (5 points a day for up to 7 days, then a zero afterward). However, if a student talks to me before the deadline and explains to me their situation, I am more than happy to give out an extension. I try my best to be mindful of all the scenarios that happen throughout a semester, and as long as there is open communication between the students and myself, this creates a positive learning environment that the students can thrive in since they do not have to fear about major penalties due to late assignments.

Speaking of late assignments, as stated previously, a mindset that I would love math instructors to break is the idea that students "know how to college" when they are a college student. No matter the age of the student, it is hard to assume that every student in our classroom will be diligent, on time with all assignments, occasionally miss class, study hard, and pass the course with an A. Wouldn't that be the life if education was like that? I bet people would be out the door trying to get a teaching job!

Now, I don't want to keep writing and have you, the reader, roll your eyes and think, "I am not here to hand-hold my students!" That is not the point. Of course, you do not need to hand-hold students; eventually, they will need to "do life" on their own and have bosses and deadlines to please and meet. But everyone at some point in their life needs a little help. Whether a student is sick, a family member passed away, or whatever other situations arise throughout the semester, let's be caring and kind to people in these situations and know we are all human. Sometimes, throwing a little rope to a student can have rewarding consequences on their performance.

Let's look at an example of this. This student will be named Ashley for the sake of the example, though this is a real story. Ashley was in an online course and struggled early by not completing assignments on time. Eventually, these assignments turned into zeros, which caused her grade to plummet to an F. I submitted a TEAMS alert on her.

As a side note, TEAMS is an online program that instructors at CGTC (and other technical colleges in the Technical College System of Georgia) can use to alert students about failing grades. An instructor submits the student's name, the class, and the reason they are submitting the alert (usually it is because of failing grades, but it can be for a multitude of different reasons). A TEAMS member will call and follow up with an email about how a student can get on track with the course. After 2 weeks, a follow-up email is sent to the instructor to see if the student has done work. If not, the instructor can choose to keep the alert open so the student is contacted again. A TEAMS member will also call to remind students of withdraw dates in case they want to take a W in the course. It is a great program since a student will receive help from not only the instructor but the college as well. The hope is that the student will become motivated enough to start turning in assignments on time and with a passing score. If your college does not have such a system, I would recommend trying to see if one can be established. This way, instructors have more than just one avenue (themselves) to get a student to do work.

Now, back to the story. After I submitted a TEAMS alert, the student sent me the following email. I did not edit the email for grammatical errors.

Hey Mr.Youmans I wanted to let you know that I did not intentionally put my self in a position to fail this first part of the class I have had a family member pass and a lot has happened since and I just wanted to let you know that I do know what to do in class I just had a family member passed and I moved out all in a week so I didn't have time to think about school. I am not asking for a re-try because I know this is college but i just wanted you to not be concerned with my learning abilities.

My response, without the formal greeting stating her name:

I apologize for the loss in your family, mental health is more important since you cannot do well in school if your mental health is not in a great place. Losing a family member is tough.

The good news is you can always turn in assignments late. If you can turn in your work for module 1 before Thursday the 9th, I will wipe away your late penalties. After that though, I have to keep to late penalties to be fair to other students.

Module 2 assignments will be due September 19th.

I first wanted to tell her that I agreed with her decision because I wanted her to know that it is okay to set aside school for life events. As I said, mental health is of utmost important, and it is difficult to do well in school if your mental health is poor. Then, I wanted to give her a chance to succeed while in the meantime being fair to the other students in the class. So, I offered no late penalties if the work is done by a certain date. Unfortunately, after this, she did not do the work but a month later sent me the following message, again without editing.

> Hi Mr. Youmans I am contacting you behave of my horrible grade nothing from your doing and only to mine. I am trying to get on top of things after I had a spiral of life event happen all really fast and then fell into a deep depression which has only caused me harm. I understand that many people go through this and understand if you don't have an advice or extra work to help me. I am in counseling and they have advised me to try and get as much a i can done and contact the teachers and ask if there is any extra assignments or way I can help get a passing grade. If not I respect your decision and will continue to try and better my grade.

My response:

> Thank you for letting me know. I know what you are going through is tough, but remember, everyone at the school is here to support you and wants you to succeed.
>
> As for my class, I am not allowed to give extra work, but I can reopen modules for you to complete. It is going to be a tight schedule, but I will allow you to remake up work. Here is the schedule that will need to be followed the catch up. . . . I know this is a tight schedule, but to be fair to other students and to get you done by the end of the semester, this is the one that we have to stick to. Remember, there is the ASC [Academic Success Center] you can always go to get help. You can get started immediately.
>
> If you have any questions, please let me know.

I intentionally left out the schedule to not bore you with those minute details. This was a tough situation for me; how do I be fair to both her and the other students who are in the class that did the work on time and successfully? I will never give free points out to a student; that would be a violation of many ethical dilemmas regarding teaching. So what I always try to do is put it back on the student. In this situation, it took me about 10 minutes maximum to reopen assignments. It doesn't take me super long to grade assignments; homework assignments are graded automatically, and quizzes and exams I can usually get through fairly quickly. So, there is no real reason for me to not let her at least try to pass the course, even if she was lying about her situation. I also do not ask for documentation myself, yes that might be asking for trouble at times, but as I said a million times, my goal is for the student to succeed, and if that is to reopen assignments because she had a rough semester due to unforeseen circumstance, I will gladly do it.

Now, this is where I see the "university effect" kicking in. Instead of being empathetic and willing to help out this student as best as I thought I could, some instructors have the mindset that this student should have been aware in the first place they had classes to complete. This student should have taken the initiative to either withdraw from the course or complete assignments in a successful and timely manner. And maybe they are right. This student, however, came to me and opened up about how her semester was going so far, so while yes, she should have been more aware of her coursework, why should I be another reason her semester has been one to forget? Instead, be the one to make a difference in a student's life by giving them a second, and heck, even third or a fourth, chance on assignments because life does get in the way sometimes, and that is okay. We as a college community are here to lift you up and help you push past this rough patch in your life. Then, give you the tools necessary to succeed in not just the class you are struggling with but all future classes as well.

Unfortunately, this student did not pass my course and received a D. However, she received an F in all other courses that semester. Here is the interesting thing: on the final exam, she received a 90%. The one reason she did not complete the course successfully was because there was just too much to overcome in terms of missing assignments. The next semester, although not doing well in all her classes, she stayed at the college and is still taking classes. Is that because of what I did for her and being kind and considerate of her situation? I have no idea. However, I would like to think that extending an olive branch to our students who are in trouble with whatever situation that is happening to them will help them that we are here to help them be successful in their path to graduation.

I do want to make a note that not all the time is extending due dates the correct thing to do. We need to be careful about what exactly it means to "extend the olive branch" to our students because sometimes extending that branch might put them at an even bigger disadvantage than if we let them redo assignments. For example, if a student is asking to reopen assignments late in the semester that could be disadvantageous for them and really a waste of time to even try. For example, I had a student who, less than a week before the semester ended, sent me this email (without me editing it for grammatical errors):

> the course is coming to an end, and I haven't used my time wisely, I was wonder would you be willing to open all assignments that I have zeros on or low grades or would you be willing to have an 'i' for a grade to let me finish any low grade to a passing grade?

My response:

> I'm sorry, but it is too late to open up all assignments. I stated in an October 18th email to complete Modules 5–7 and the final, but assignments from Module 6 and Module 7 were not completed and both the quizzes were not passed as well. So, even giving an incomplete grade would not help.

Saying no to this request was difficult, even though I knew that reopening all assignments would be fruitless based on the work that had already been submitted by the student. The student realized a little too late that time was not spent well, and while I may have been more empathetic a month or two before she sent this email, right at the end of the semester, it would not be fair to her or the other students for me to open it.

Honestly, it is a fine line between "How can I help you out?" versus "I'm sorry, you will have to take the course next semester." Really, it is a case-by-case basis. I ask, however, is that we as instructors give each case a fair chance so that we can put students in a place to succeed. Reaching out to students, seeing how we can help them, and following through with guidance and recommendations on how to pass is a great way to start making an impact on a student's life.

Coming from a university, sitting in 100-person classrooms, I could understand why instructors have stricter deadlines and have a harder time "extending the olive branch." Having upward to 500 students a semester would make life difficult if trying to please everyone. However, let's not kid ourselves, most technical college faculty are not seeing 500 students a semester. At CGTC, faculty normally see 100 to 150 students a semester. In the fall of 2021, I personally had 49 students enrolled (although attendance was low) in face-to-face classes and 51 students online for a total of 100 students. So, I really see no reason why class sizes should be a factor in trying to help students throughout a semester.

Overall, let us be kind and sensible instructors, especially in mathematics where students are already scared and anxious to be in it. As I have said before, confidence is half the battle when it comes to a student's chance of success in a mathematics course. A student's confidence will grow if we treat them as humans and tell them that it is okay to not be okay throughout the semester. The big idea, I tell my students, is to make sure to communicate with me what is going on so that I can help them out as best as I can. I am not a mind-reader. Things happen, whether being stuck on a problem or getting sick and not being able to complete assignments.

Whatever the case may be, let's try to show compassion to our students so they can graduate and get the career of their dreams, the ultimate sign of perseverance. In turn, our mathematics classrooms will become a place of learning, support, and growth. That, in my opinion, is how mathematics should be—not a subject to be feared or loathed but a subject that can promote critical thinking, time management, and perseverance. Hopefully, by the end of their college careers, students will look back and see that their math course was one of the best.

REFERENCES

1. *Utah State Fast Facts.* Utah State University. www.usu.edu/about/fast-facts/
2. *USU Demographics & Diversity Report.* CollegeFactual.com. www.collegefactual.com/colleges/utah-state-university/student-life/diversity/
3. *Data USA: Central Georgia Technical College.* Datawheel. https://datausa.io/profile/university/central-georgia-technical-college
4. Rutschow, E. (2019). *The National Academies of Sciences, Engineering, and Medicine Workshop on Understanding Success and Failure of Students in Developmental Mathematics: Developmental Mathematics Reforms.* Workshop on Understanding Success and Failure of Students in Developmental Mathematics. https://sites.nationalacademies.org/cs/groups/dbassesite/documents/webpage/dbasse_191791.pdf
5. *Mission Statement.* Utah State University. www.usu.edu/president/missionstatement
6. *History, Mission, Vision, and Values.* Central Georgia Technical College. www.centralgatech.edu/about-cgtc/history-mission-vision-and-philosophy

CHAPTER 4

The Application Conundrum

Without mathematics, there's nothing you can do.

—SHAKUNTALA DEVI

One of my favorite opening lines of a TV show is from the show *Numbers*.

> We all use math every day. To predict weather, to tell time, to handle money. Math is more than formulas and equations. It's logic; it's rationality. It's using your mind to solve the biggest mysteries we know.

Funny thing is I wanted to be a meteorologist before becoming a math instructor. I guess I wasn't a big fan of being wrong almost all the time (just kidding, meteorologists every day do lifesaving work).

Applications (word problems, as my students would say) are one of the more dreaded topics we discuss in any of our math classes. I gave my students a final exam in class during the fall of 2021, and I gave my students the option of picking 8 questions out of the 11 given on the exam. As one student said, "I am not doing any of the word problems!" The good news for her was that there was only one word problem, and the rest were solving questions!

DOI: 10.1201/9781003287315-4

The interesting part of math courses at technical colleges, however, is that students *want* to see applications of mathematics. How many times have we heard the simple phrase "Why do we teach algebra in high school and not how to balance a checkbook?" Ugh. The same sentiment happens at technical colleges. If a student is trying to get into the nursing program, why do they need to learn about how to use elimination to solve a system of equations?

So, this is why I call this chapter the application conundrum. On one side of the ring (wrestling ring, not an algebra ring; bad joke, sorry) are applications of algebra. Students yearn for them. Students want to see why solving for *x* is actually important. On the other side of the ring is a very similar opponent, word problems! Students hate them. Students want to skip them. They get upset when one example is done in class, and the example on the homework looks completely different! Who will win? No one knows! My money is on students knocking both out.

Before getting into how on earth we can talk about these applications without students becoming anxious at the mere thought of solving word problems, I want to talk a little bit more about what I believe is a bigger concern, and that is the hatred of mathematics in general. I say a little bit more because I have and will continue to discuss how we can overcome these barriers throughout this book, but I wanted to take the time now to discuss something that is also applicable to this chapter, and that is the fact that student's fail to realize that people do love mathematics, including the instructor that is standing in front of them.

Now, I don't want this to be a pity party. Oh, poor me, my students hate the subject that I teach. That is not necessarily what I am saying, although a party does sound nice as long as there is good food. Maybe a taco bar.

However, I do want to stress that sometimes it gets difficult with people hating on a subject you've spent years studying. Teachers all the time get thrown in the deep end due to their profession, and truly it isn't fair to them. Questions on curriculum, books, grading, and so much more has teachers burning out of the profession at astonishing rates. These rates are discussed further in Chapter 7, so I won't get too much into this conversation right now.

Teaching, and more especially math teaching, is something I, and a lot of others, love to do. Not only that, but there are also so many career fields that use mathematics; in fact, almost every single one does without a student realizing. Like I said before, no that does not mean they will be using elimination to solve a system of equations, but that doesn't mean *someone* isn't using that skill set in their job. That also doesn't mean there isn't an application out there for students that are in the class not becoming a mathematician, but with so many lessons needing to be completed throughout the semester, it is difficult to find the time to cater to all the different applications students want to see applied to their own careers.

So, how do we cater to those in a class students from a diverse range of majors? There are a few possibilities.

1. Create word problems dealing with everyday life, not necessarily catered toward specific jobs.

Here is a riddle for you: what is one thing in the entire world that people love, they use every day, but can cause heartache depending on how it is used? That's right! Money is the correct answer, here.

If it is difficult in your class to teach word problems based on the job fields of students in the class, let's teach word problems a little more practically. If we create word problems based more on how students can use them in the real world, then they will start to see the usefulness of the mathematics being taught.

Many applications in College Algebra can be solely focused on one thing: money. Now, I am not saying to make *all* your word problems about money, since that would get boring for the students. But if there is anything that will get students excited about learning mathematics, it is money.

Take, for example, lessons on exponential and logarithmic functions. These topics are pretty typical for a College Algebra course taken at a college. One application of exponential functions is interest rates, which compounding interest can be calculated using the following formula:

$$A(t) = P\left(1 + \frac{r}{n}\right)^{nt}.$$

Compounding interest is interest earned on both the principal and the interest already earned. In this formula, $A(t)$ is the amount earned after t years, P is the principal (the original amount), r is the rate (usually expressed as a percentage), and t is time. Typical problems that are given out state things such as:

If I were to put $10,000 into a bank account with a 3% interest rate compounded semiannually, how much money would I have in the account after 5 years?

Yes, this in itself is a great problem to see why the mathematics we are doing is important. Typically in class I try not to do an example such as this. Why? Because they get to see three or four examples on the homework. Instead, I take the opportunity to talk to students about retirement. I always try to pick on the youngest in the class and ask them how much they want to have in a bank account for retirement. They usually say something along the lines of $100,000, and I look at them and say, "You only want $100,000 in retirement?" Usually it gets a good laugh.

Anyway, we work the same type of problem, but instead, I express it this way:

> Student A wants to save $1,000,000 for retirement. Student A puts $1,000 into a savings account that compounds quarterly at a 3% interest rate. How long will it take Student A to retire with $1,000,000, assuming no additional deposits are made into the account?

If you are curious, the answer is approximately 231 years. Students get a good laugh out of this, because it is ridiculous to think that you can save that much in 231 years! This can then lead to a discussion about Roth IRAs, 401k, and other forms of retirement that even 18-year-olds can start to think about. I even had an older student of mine state that her grandchild invested in cryptocurrency during this lesson, which perked up the ears of some of my younger students in class. If only we all invested in Bitcoin 10 years ago! I guess now they—whoever they are—can't say we don't teach relevant information to our students in a College Algebra course!

We can take applications of money even earlier in the course than exponentials and logarithms, since typically those topics are reserved for later in the class. Even going to linear equations in one variable, we can ask our students questions regarding tax and simple interest rates. For example,

> If I bought a car for $10,000, but my total came to $15,000, how much tax was paid?

"Five thousand dollars!" one might scream. True! This example can be formed into a (linear) mathematical equation and then solved for the variable. So, money problems are a great starter for word problems if students are struggling to see why the mathematics we are doing in the course is applicable to them.

A side note for this, however; instructors do need to be careful about, one, using the textbook for the course to come up with word problems and, two, presenting the same type of problem to students that will eventually not require critical thinking. Julia Gattuso at the University of Mary Washington put it best in saying that "research shows that textbooks do not provide students with the necessary skills needed to solve word problems" because "they present word problems in such a way that once a student learns how to solve one problem . . . they automatically know how to solve the rest of the problems . . . because they are plugging in the same formula for each problem."[1]

It's not beneficial to students if they are working problems in the same exact manner every time, and that is something I pointed out to begin this chapter is that students complain that on the homework, all the word problems are different in nature. But we want to develop *critical thinking* in our students. For example, how can we apply a problem about money to a problem about mixing chemicals to yield a desired result? This is where it gets difficult for a math instructor at a technical college, however, and that is we must balance the line between critical thinking word problems and the mission of the technical college, which is to help students succeed and graduate in a timely manner. If students are coming into our classes wanting to graduate fast, do we spend the time developing deep critical thinking skills, or spend just enough time to say we completed the standards set forth by the state and that they can develop more skills when they are in their major courses?

This is a difficult question to answer, and I think it truly depends on the instructor and what exactly other departments in the college are looking for when they are selecting students for their programs. Here at CGTC, our homework relies on critical thinking skills. They get to use the skills that are taught in class on how to solve word problems and then use them for a diverse group of problems. This diverse group usually includes money problems, distance-equals-rate-times-time problems, and mixture problems. This, in turn, will help students with their critical thinking skills so that when they get to their major courses, they are ready for the rigor that those courses provide.

2. Have students create word problems based on their own experiences and have other students in the class solve them.

I think this idea is a great way for students to show off what they love to do to other students in the class, and then they also get to see how other students will be using the mathematics we talk about in their majors as well. The first part of this project requires students to think about how they will be using mathematics in their everyday lives and careers. You can't create a good problem if you don't know how it would fit in with your career! The second part has them actually solve other problems students create so they not only get the problem-solving skills necessary to move on the course, but they also get to see what other students are doing as well. It's a win–win.

I have tried this in the past, and I offered it as an extra credit opportunity for my students one day toward the end of class. I had just moved to Georgia, and it was an absolutely gorgeous day outside. This was in late February, and if you have ever been to Utah, you would know that the weather in late February is a lot nicer in Georgia than it is in Utah (unless you like snow). So, I decided to take the class outside and have them form pairs to create word problems that other students would solve in the class. They used the world around them to create these problems. I think it was a great activity for them to see how the world around us, just like what *Numbers* says in the introduction, is filled with mathematical concepts.

3. Create projects for students that not only deal with the mathematics being taught about but are somewhat fun in nature as well.

"Fun" is a very relative word here. What is fun for me as the instructor may or may not be fun for the students, especially when the project is for a grade. However, we want to create projects for students that have them not only develop critical thinking skills but that can keep the students engaged as well. Now, I am not going to claim I am good at making "fun" projects for students; I have a long way to go in project development. But I am quite proud of one project I had students do, so I want to share that with you.

At Utah State University, the College Algebra course also taught on exponentials and logarithms. One such application of exponential equations is growth and decay. So, the project setup was this: the world population is predicted to be 9.8 billion by the year 2050.[2] Countries need to model the population growth for their country so know how to provide for the people. The job of the students, in groups, was to pick a country and research the different questions that I asked of them. Then, they created a PowerPoint presentation of the information they found and presented their findings to the class. The following were those questions:

- Present a brief history of your country (location, language, etc.). Every group had a different country.

- Use the Logistics growth model on page 307 example 9 (the text-book used in this course was *Precalculus: A Unit Circle Approach*, 3rd edition, written by Ratti, McWaters, and Skyzypek) to calculate the average rate of the population of your country given the popula-tion of your country starting at year 1950 and using the year 2010 in your calculations. They had to show their work in the PowerPoint and in the talk explain why you chose the number for the carrying capacity.

- Use the exponential growth model to predict both (1) the population of the country the year 2050 and (2) predict when the country will reach x, where x is an number of people higher than the number of people in year 2010. The k in these formulas (the formulas were pre-sented in the textbook) you can use as the percentage you calculated in the first item.

- Create a graphical representation of your data, these data being the years from 1950 (or a year close to it in which you have the data) to 2050 in increments of 10 years.

- Explain using your graph how your graph represents an exponential model.

- How would the population increase affect your country?

- Create a 1- to 2-page plan that will help your country in mitigating the affects you said in the previous step.

- Oh no! A disaster has hit your country in the year 2050, and the pop-ulation is now decreasing. Assume the rate of decrease is the negative of the rate of increase you calculated in step 1. First of all, what disas-ter has hit? What year will your country reach a population of 0?

Feel free to steal this project or change it to fit your needs. Let me explain what my thought process was on this project and why I thought it would be "fun" for the students.

First, the scenario that was created seems like a real one, not one that was arbitrarily created. Application questions are often designed without thought to if the question makes sense in the real world. For example, one application question asked to our College Algebra students during the linear equations in one variable section is the following:

> John painted his most famous work, in his country, in 1930 on composition board with a perimeter of 105.39 inches. If the rectangular painting is 5.51 inches taller than it is wide, find the dimensions of the painting.

While this is a great question to assess students' knowledge on both picking out pieces of information to create a mathematical equation and then solving that equation for the desired information, the problem itself does not seem very realistic. Why would anyone care about finding the dimensions of the painting in this matter when you could use a ruler to find how tall and wide the painting is? Two questions later this question is asked, which I think is more applicable to the lives of students:

> A hotel manager found that his gross receipts for the day, including 6% sales tax, were $4,358.72. Find the amount of sales tax collected.

While students in the class might not be hotel managers when they graduate, the usefulness of finding the amount of tax that was collected during the day would be useful since tax is usually given back to the government. So, whether it is at a grocery store or a hotel, finding the amount of sales tax is important, and this problem can still be solved using a linear equation in one variable. So, giving problems that seem to be realistic (and I believe the project I gave was) will motivate students to perform at their best during the project.

Next, I wanted to incorporate other subjects that students might be taking as well, in this case history. Students, even at technical colleges, often need to take some form of a history or humanities class, so trying to incorporate a project in a mathematics class that ties in with their other courses will let them see the different connections between classes. For example, at the University of Northern Iowa, I took a class specifically on the history and culture of India for a humanities credit. If I was doing this project at that time, I could tie in what I found out about India in that class (and one of those is that India is a very populated country) to see what over-population might do to that country. I also had them do a writing assignment (creating the 1- to 2-page plan that will help the country mitigate the effects of overpopulation) that would tie in with any English class nicely. While this may not be backing up my point about it being a "fun" project, I still view the importance of general education courses and how they all tie together to be an important one.

Third, to hopefully create a "fun" scenario for the students, I had them think about the negative rate of change for the country if a disaster hit their country. They got to create their own disasters, which unfortunately most of them picked a disease that killed most of the population. This somewhat backfired on me, since I wanted students to use a little creativity here. Maybe a volcano erupted in their country, forcing everyone to evacuate. Or maybe a dinosaur was hatched and forced everyone to evacuate the country and a wall was put around the country to maintain the dinosaur. However, most everyone picked a plague hitting the country. Oh well, I guess I tried!

Again, I'm not going to claim I am the best at project making. Maybe this was a bad project to give them, and if that is your opinion, I respect that. However, I wanted to create a project for students to complete that felt real. In the two classes I did this project in, students seemed to enjoy it as long as the group they were with was a productive group. There are many studies on picking out groups in classes, so I won't talk about those strategies here. Overall, however, the project was a success and students got to see an example of how the mathematics done in class comes to life.

4. Teach toward the homework.

Oh, a controversial one! I can already read the emails I will get for putting this one in this chapter; teaching to the homework is not promoting good critical thinking skills, and they really aren't learning anything by just teaching directly to the homework. I would like to make an argument that teaching directly to the homework can be beneficial up to a point.

During the CGTC College Algebra course, there are five specifically designed lessons that only have word problems in them. I talk about this later in Chapter 7 as well, but before I started directly teaching from the homework, I was just writing down problems on the board to complete. Students would then in turn go to the homework and not perform successfully on them because they would think that the problems I did in class were nothing like those in the homework. This is where I can see the argument of critical thinking skills coming into play for *not* teaching toward the homework. Lecture is supposed to give you the skills necessary to go to the homework and reason your way through problems, even if the problem doesn't look exactly like the one did in class.

However, I wanted to change that strategy to teaching problems that were specifically on the MyMath Lab assignment for two reasons, one because then students cannot say that we did not do problems similar to those on the homework, and two because MyMath Lab, while the problem still looks similar, will change the numbers so the problem is not exactly the same. They will have to think about what we did in class and apply it to the similar but still new enough problem. You will see more specific numbers in Chapter 7, but for now, the scores on the first homework assignment raised by an average of 5.7%. Not a ton, honestly, but even a little bit of positive change will go a long way toward student success.

There is one other reason why I wanted to start teaching toward the homework for the application sections. That reason is that we are using technology (MyMath Lab) to deliver the homework assignments. Since one of the first assignments is on applications of linear equations, it gives me a good opportunity to go over the software and the usefulness of it. My apologies if you do not use MyMath Lab or similar software in your class; you can probably skip ahead a little bit here! I can show them how to input answers and how to utilize some of the extra help that Pearson's MyMath Lab provides. For example, two buttons that are often used by students are the "View an Example" button and "Help Me Solve This" button. It also gives me a good opportunity to explain to students that while these buttons are helpful, sometimes the software teaches the student a slightly different method to solve the problem than how I did it in class. Now, I am an instructor who does not live by the mantra "my way or the highway." I talk more about this idea in Chapter 5, but for now, it is important for students to realize that my way, or MyMath Lab's way, is not the only way to solve a problem.

So, teaching toward the homework is, and maybe rightfully so, looked down up. We should be teaching our students the critical thinking skills necessary to succeed in not only the mathematics course they are taking but also in their future major classes and career as well. However, I don't think it is necessarily a bad thing to teach toward the homework in certain situations, including application problems. They can still get similar but new problems by switching up the numbers in the problem. Then, applying what they learned in class is key to successfully completing that section's homework.

However, an instructor teaches about applications, it is important to let students see applications of what they are doing in class. I try to remind my students that historically, College Algebra is a course that prepares students for calculus, so not all topics in College Algebra will be applicable to a real-life event. There isn't anything wrong with a topic not being relevant to your life. The key is to understand that we are helping our students develop skills, specifically critical thinking skills that will be necessary for their major classes and careers. If the goal of a college is to develop a well-rounded citizen, these critical thinking skills gained through word problems is a great way to achieve that goal.

Overall, applications can be fun and relevant to a broad range of students. If we take some time and create examples in our classes that speak to the diversity of students in our classes, that will in turn create a sense of inclusion for our students. They will be able to see the relevance of such applications and, in turn, develop necessary critical thinking skills to help them succeed later on down the road. Maybe, just maybe, they will start to like mathematics as well and tell their friends all about how mathematics can be used in our lives when they question it as well. A man can dream, I suppose!

REFERENCES

1. Gattuso, Julia. (2015). *Problem Solving Strategies: Helping Students Develop a Conceptual Understanding of Word Problems*. Student Research Submissions. 182. https://scholar.umw.edu/student_research/182
2. *World Population to Reach 9.9 Billion by 2050*. International Institute for Sustainable Development. https://sdg.iisd.org/news/world-population-to-reach-9-9-billion-by-2050/

CHAPTER 5

The Rigor of Mathematics

In mathematics, rigor is not everything, but without it, there would be nothing.

—HENRI POINCARÉ

Mathematicians reading this book know that our subject can be difficult because of the rigor behind it. One seemingly innocent topic in mathematics still needs to be studied deeply and each conjecture needs to have a proof before calling it a theorem. I remember learning about the ins and outs of proof writing and just how rigorous mathematics can be in a class called DAM. "This DAM math class," students used to say while taking the class. Not only did I learn how to write proofs in Discrete and Argumentative Mathematics, but I also learned to argue through logic, present mathematics to a group of people, and teach in front of a classroom full of engaged learners.

Fun fact: One other thing I learned in this class is when you write on the board and want to erase, do you erase up or down or left to right? The answer is that you should erase up and down! The reason? If you erase left to right, you shake your behind at the class, and who really wants that?

As one moves through higher-level math courses such as analysis, number theory, abstract algebra, and much more, they start to dive into the beautifulness of mathematics. That is, the beauty of why topics we teach

DOI: 10.1201/9781003287315-5

in high school actually are cared about. Let's take a look at an example of how mathematics can take a turn toward being rigorous.

THEOREM: ADDING AN ODD INTEGER WITH AN EVEN INTEGER RESULTS IN AN ODD INTEGER

I would consider this a number theory problem. First, let's talk about the wording of the theorem. This seemingly trivial example presented to a group of students taking a mathematics class at a technical college may have them confused. Most students will know what addition is. However, what is an integer? As the instructor, you will know that an integer is the set

$$\{..., -3, -2, -1, 0, 1, 2, 3, ...\}.$$

But to a group of students among which some may have not taken a mathematics class in 20 years? If they haven't taken a class in a long time, maybe the words *even* and *odd* are both mysteries to the student. These words may have been forgotten or were never taught in the first place. So, before you assume that students know what all parts of the theorem say, make sure to explain each piece so everyone is on the same page. We can even focus on using the word *theorem* to this demographic of students. Yes, as mathematicians, we know that theorems drive mathematics (well, axioms do, but that can be a conversation for another day). If a statement can be proved, we should call it a theorem. Should we use that terminology with our students in technical education? It depends on the class. In a College Algebra class, as long as you explain yourself, then probably. If you are teaching a Foundations of Math course in which you are teaching how to add numbers, then I most certainly would not. Especially in college algebra–type classes, let's not be scared to introduce our students to the wonders of mathematics, and part of that wonder is, in fact, looking at different theorems and wondering why they are true. Here is one way you could introduce the theorem about even and odd integers to your class.

"Okay, class, today we are talking about the addition of numbers. Recall that an odd number is one that is not divisible by 2 and that even numbers are those that are divisible by 2. Now, what happens if we add two numbers, say, 3 and 4, together? Would we get an even or odd number? Well, 3 + 4 = 7, which is an odd number. Write down an even and an odd number on your paper and see if the result is even or odd. It should be odd, no matter what you right down! This leads us to think that the following statement:

An even number added to an odd number results in an odd number.

Notice that I did not even say the word *theorem*. You can if you want. As long as you have already introduced the idea of what a theorem is to your class, then go for it. There is no need, however, to students who are learning about adding numbers together. Also, notice I left out the word *integer*. Especially to a group of students in a developmental math course, we do not need to introduce terminology that will scare them. Let them *understand* the topic first, and then slowly start to introduce the terminology. Now, having students look at a proof of your statements is also an interesting debate to have. For the purpose of the example, let's look at a proof of the statement that an even number and an odd number add together to give an odd number. Let's remind ourselves of the theorem one more time.

Theorem 1: Adding an Odd Integer to an Even Integer Results in an Odd Integer

Proof: Let k be an integer. By definition, $2k$ is an even integer, and $2k + 1$ is an odd integer. Then,

$$(2k) + (2k + 1) \, 4k + 1.$$

$4k + 1$ is an odd integer, thus proving the theorem.

This is mathematics at work. We took a seemingly innocent statement and proved the result using definitions that would have been taught in a previous lecture, and now we know that the statement will hold for any even integer added together with an odd integer.

Let's think about Calculus I as well, since this type of class can also be taught at technical colleges. When you first introduce the derivative, the proper way is to have them understand the topic of instantaneous rate of change through the use of limits. Then, you slowly start to introduce them to rules such as the power rule, chain rule, and product rule. There are reasons why you do it this way, first off, *calculus is limits*. Everything you do in calculus is about limits, derivatives, integrals, sequences and series, and so on. You should not just jump in and teach them about the product rule before having them understand *why* you are taking derivatives in the first place! Second, if you truly want to see why the product rule is actually a true statement (or really any of the derivatives rules for that matter), you need to use limits to prove the statement. Take a look!

> **Theorem 2: Let f(x) and g(x) be continuous, differentiable functions. Then,**

$$\frac{d}{dx}(f(x)g(x))=f(x)g'(x)+g(x)f'(x).$$

Proof: Suppose that $f(x)$ and $g(x)$ are continuous, differentiable functions. We want to show that

$$\frac{d}{dx}(f(x)g(x))=f'(x)g(x)+g(x)f'(x).$$

Since f and g are differentiable,

$$\lim_{h\to 0}\frac{f(x+h)-f(x)}{h}\text{ and }\lim_{h\to 0}\frac{g(x+h)-g(x)}{h}$$

exist. By continuity and the definition of the derivative for $f(x)g(x)$,

$$\lim_{h\to 0}\frac{f(x+h)g(x+h)-f(x)g(x)}{h}$$

$$=\lim_{h\to 0}\frac{f(x+h)g(x+h)-f(x)g(x+h)+f(x)g(x+h)-f(x)g(x)}{h}$$

$$=\lim_{h\to 0}\frac{f(x+h)g(x+h)-f(x)g(x+h)}{h}+\lim_{h\to 0}\frac{f(x)g(x+h)-f(x)g(x)}{h}$$

$$=\lim_{h\to 0}g(x+h)\lim_{h\to 0}\frac{f(x+h)-f(x+h)}{h}+\lim_{h\to 0}f(x)\lim_{h\to 0}\frac{g(x+h)-g(x)}{h}$$

$$=g(x)f'(x)+f(x)g'(x).$$

Phew, we did it. Students at the university level may see this and get excited. "Yes, we did it!" they would shout. Well, that is my dream anyway. When I actually presented this to a Calculus I class at Utah State University, they all stared for a bit wondering if they will have to do the same for the upcoming exam. They might also be inspired that until this theorem is proved, it cannot be used, and the limit definition of the derivative should be and must be used to solve problems instead. I haven't met a student who wants to do that.

For students at technical colleges who do not have a background in mathematics, is the proof necessary to talk about how to find derivatives? In my opinion, yes. Sometimes, rigor is needed no matter the background of the student, and Calculus I is a great place to start introducing the ideas of proofs and theorems. Also, if a student was to take a Calculus II course at a 4-year university, the rigor in that class would need to be introduced in a Calculus I course.

While theorems 1 and 2 are fantastic, students often do not move on to Calculus I or number theory at a technical college. The majority of students are taking a College Algebra course. So, let's examine a College Algebra class. An often-taught section in college algebra is multiplying polynomials. Generally, FOILing—first, outer, inner, last—is taught. Really, FOIL is just an application of the distributive property: if a, b, and c are real numbers, then

$$a(b+c)=ab+ac$$

But there are some special rules you can teach students to make multiplication a little quicker, and they are helpful when you get to the section on factoring. The three I teach are the following:

I. $(a+b)(a-b)=a^2-b^2$

II. $(a+b)^2 = a^2 +2ab+b^2$

III. $(a-b)^2 = a^2 -2ab+b^2$

This, while seemingly looking innocent, is a beautiful way to get your students to start thinking about *why* such formulas are true instead of just giving them the formula and doing examples. We just learned how to FOIL, so now FOIL $(a + b)(a - b)$:

$$(a+b)(a-b)=a^2 -ab+ba-b^2$$

Notice that $-ab+ba=0$. This is one way the rigor of mathematics could get in the way of the learning of beginner students of mathematics. To be rigorous, it would be ideal to state something along the lines of "Since a and b are real numbers, they are commutative so $ab = ba$; therefore, $-ab + ba = -ba + ba = 0$." This is not needed for students who mathematics is not their major, I personally am content if they were to write

$$(a-b)(a+b) = a^2 + ab - ba + b^2 = a^2 - b^2.$$

The other two rules can be discovered the same way. Have the student FOIL out part II and part III, and they would get, respectively,

$$(a+b)(a+b) = a^2 + ab + ab + b^2 = a^2 + 2ab + b^2$$

and

$$(a-b)(a-b) = a^2 - ab - ab + b^2 = a^2 - 2ab + b^2.$$

This I would consider beautiful work. They took a concept (FOIL) and applied it to a difference of squares (the name of part I) to see how to multiply two polynomials that are a difference of squares. Then, to really solidify the concept, have them work an example such as find the product of $(2x+3)(2x-3)$. Then, here comes another interesting take on how we as instructors perceive student work. Would we be happy if they just FOIL out the two binomials, or do we really want them using part I? For students who are taking a College Algebra course for general education requirements, I see no difference between the two. If they show their work and get the right answer, then we should celebrate their achievement no matter how they got it (unless, of course, the work itself is incorrect, and they got lucky in the final answer).

But here is the funny part, in a Calculus I class, I would be the complete opposite of "as long as they get the answer, I am happy." If I am teaching on the limit definition of the derivative (or, for that matter, the limit definition of the integral), I would not want a student using the power rule for derivatives even though that is a correct way of getting the correct answer. Why? For one, at the point of teaching the limit definition, I had not taught on the power rule, so it would not be fair to other students if I allow students who have taken the class before or already know about the power rule to bypass the limit definition. Also, as previously stated, I have not proved the power rule at that point in the class, so based on the rigorous nature of mathematics, I should not be allowed to use the power rule until that theorem is proved. Finally, while the power rule is perceived more easily than that of the limit definition, some examples that can be done in Calculus I cannot be done with any of the rules except for the limit definition. One example of homework I have given my students before is

> Give and explain an example of a function that is first differentiable but not second. (Hint: Consider a piecewise-defined function.)

This example requires the use of the limit definition. If you want to know, a great example is

$$f(x) = \begin{cases} x\sin\left(\dfrac{1}{x}\right), x \neq 0 \\ 0, x = 0 \end{cases}.$$

Away from zero, everything is fantastic. All derivative rules can apply. However, at zero, all limits are off (pun intended). You have to use the limit definition to show that the derivative exists at $x = 0$. The two semesters I gave this problem out, there wasn't too much success with this problem without guidance from me.

In a college algebra–type course, there are analogous topics that the idea of "my way or the highway" might need to happen. Take a look at factoring and the quadratic formula. When we start solving quadratic functions using the zero-factor property (for clarity, this property states that if $ab = 0$, then $a = 0$ or $b = 0$), some students who learned the quadratic formula in past classes want to jump straight in and use it. For those who are reading that do not know about the quadratic formula, if you want to solve a quadratic function of the form $ax^2 + bx + c = 0$, then you can use the quadratic formula to find x, which this formula is

$$x = \frac{-b \pm \sqrt{b^2 - 4ac}}{2a}.$$

Yes, this can be used to solve every single quadratic equation (after, of course, proving that it can, which by the way is a fun and fantastic exercise for students after showing how to complete the square). However, it would be a great disservice to let students just use the formula for each and every example. This is where we as instructors need to know the material and realize why factoring can be another great tool for students to solve problems. Take a look at this application problem from one of the homework assignments I give out to my College Algebra students after talking about factoring and the quadratic formula:

A ball is projected upward from the ground. Its distance in feet from the ground in t seconds is given by s(t) = −16t² + 208t. At what times will the ball be 576 feet (175.6 meters) from the ground?

Since $s(t)$ is the function that gives the distance from the ground, and we want to know the times the ball will be 576 feet from the ground, we need to find

$$576 = -16t^2 + 208t.$$

Students would (hopefully) know to set this quadratic equal to 0 by subtracting 576 to both sides of the equation, which would result in

$$0 = -16t^2 + 208t - 576.$$

Students who wish to solve this by the quadratic formula are free to, but again, it is my opinion that this would be a disservice. Look at this quadratic in the quadratic formula.

$$x = \frac{-208 \pm \sqrt{208^2 - 4(-16)(-576)}}{2(-16)}$$

Bring out the calculator, because I definitely wouldn't want to solve this by hand! If you are a classroom where calculators are not allowed (I don't necessarily believe in this since our world is full of technology, and it is important to know how to use this technology to get desired answers, but that can be another topic for another day), this problem would take quite a while to complete. Instead, it can be recognized that each term in the equation is divisible by −16. If each term is divisible by −16, we can write the polynomial as

$$0 = t^2 - 13t + 36.$$

This is quite fantastic, because now it can be factored! Two numbers that multiply to give 36 but add to give me −13? How about −9 and −4? So, the quadratic can be factored into

$$0 = (t-9)(t-4).$$

Using the zero-factor property, this would tell us either $t - 9 = 0$ or $t - 4 = 0$, which adding 9 and adding 4 to both sides of each equation, respectively, results in $t = 9$ and $t = 4$.

I personally still give my students the choice between the factoring and quadratic formula once both have been learned. I disallow the quadratic formula until it is learned. Once both are learned, however, students get to pick which method is their favorite and use it. Students are learning critical thinking skills when they get to pick their method of solving. In their future jobs, there are multiple ways to get the work done which is asked. However, the boss is relying on you to know what you are doing and get the work done correctly and on time. The same argument holds here. I gave you the tools to be successful; now I am relying on you to work homework problems and get the correct answer.

Mathematical rigor can also start to be introduced at the developmental math level. While I would definitely not go to the extremes of theorems and proofs, there are topics in these courses that students can start to discover for themselves rather than the instructor giving the algorithm to solve. We can also start getting our students to start questioning *why* such algorithms are true rather than just accepting them even though the answer to why may be beyond the scope of the course.

For example, in a prealgebra course, you may teach the commutative and the associative properties of addition using integers. We can have students discover these properties on their own: give them a few numbers to play with. For example, add 2 and 3 together; then add 3 and 2 together. Do the same for -1 and 3 and 3 and -1. This is the commutative property in action, which states that if a and b are integers (really, any real number but again we are only working in integers in this example!), then $a + b = b + a$. Now, give them three numbers and ask them to add them any different orders. They will discover the associative property, which states that if a, b, and c are integers, then $a + (b + c) = (a + b) + c$.

Fantastic! So, your students have started to think about examples of the commutative and associative properties, and now it is assumed that it works for *all* integers. As a mathematician, one of the seven deadly sins is to never assume something is going to work no matter how many examples are tried. So, we should prompt our students to start asking *why* addition is commutative and associative with integers. Yes, it works for examples, but does it work all the time? Well, yes, but how we do know?

The real answer is that the set of integers is an abelian group under addition. Whoa! Hold up. This just got real. A seemingly innocent statement of integers is associative and commutative with addition now turned into throwing around words such as *group* and *abelian*? Don't worry, though, you would *never* mention this to students in a developmental math course. However, this is why mathematicians take courses such as abstract algebra because in the back of their minds, they know why addition is associative and commutative and that can help in teaching students how associativity and commutativity work. Similarly, mathematicians should take a real analysis course before stepping foot in front of a Calculus I class. Alternatively to teaching group theory, we should be happy with students starting to question if these facts hold true for all such numbers and move on to bigger and better things.

Let me go back to a statement I said earlier which I did not expand on: "the beauty of why topics we teach in high school actually are cared about." It's true that mathematical rigor such as proof writing is usually not taught at technical colleges; a lot of students struggle to even pass college algebra let alone trying to write proofs of the statements written down in the class. However, it is of my opinion one of the reasons students do not like learning about mathematics in school is because they do not get to see *why* the topics taught in high school are important to future mathematics courses.

Mathematics is very similar to the classic *A Tale of Two Cities* written by Charles Dickens. As someone who became a mathematician and not an English instructor, forgive me if I completely missed the point of the book, but for me, the book really was not a great read until the very end. It felt disjointed, and it just felt like a book I really didn't like. Until the very end. The last few chapters tied the entire book together like a knot and made the experience of reading the book worth it. Without remembering characters, storylines, or how the story ends, what I remember about the book is that I loved the book for how it started out terribly but ended in a fantastic manner.

For a lot of our students, I feel this analogy is similar. They take high school classes and learn about the average rate of change (slope), x- and y-intercepts, solving equations, solving a system of linear equations, exponential and logarithms, and whatever other topics that might be taught in a College Algebra course. However, all those topics are only half the story. They do not get the full story of how these topics can be used in further mathematics courses.

Take for example systems of linear equations. If you aren't familiar with this topic, what you are usually asked to do is solve two (or more) linear equations simultaneously for the unknown variables. For the sake of the examples, let's just stick with two unknown variables x and y and two equations. At our college, and I expect many other colleges, it is taught to solve these equations by substitution or elimination usually depending on the form the equations are given in. If they are both in the standard form $Ax + By = C$, elimination is usually picked. If one (or both) of the equations are solved for one of the variables, substitution is usually picked. After this is taught, a few application questions are looked at such as distance equal rate multiplied by time problems, chemistry problems, and the typical "If two towers have a height of 300 feet and one is 200 feet taller than the other, find the height of both towers"–type questions. While I talked about applications in Chapter 4, it is my belief, from experience, that students, even after doing some applications, do not appreciate the beauty of *why* systems of equations are important.

Of course, I am being naïve in the sense of assuming students who are in trade-type programs would think seeing if two vectors form a basis is beautiful. However, one of the reasons why solving a system of equations is important is because it helps when you learn in a Linear Algebra course that two vectors form a basis in $\mathbf{R}^2$ if they are linearly independent and span the set. You can put the coefficients of the system into a matrix and then perform row operations in them (which is actually a hidden way of doing elimination and substitution) to see if one of the rows turns up all zeros or not. If a row is all zeros, the vectors are not linearly independent; therefore, they are not a basis. Again, you would probably not discuss vectors and bases in a college algebra class or any class at the technical college level for the matter. As a mathematician who loves linear algebra (and what mathematician shouldn't? It drives a lot of mathematics!), students just see the practical use of solving systems of equations and not really seeing why it is useful for the future.

Another interesting example of this might be a little more relevant since Calculus I is taught at technical colleges while Linear Algebra usually is not, and that is average rate of change. Average rate of change, or slope, is an important concept taught because the applications are boundless. However, so many more wonderful applications can be found when going one step further and talking about *instantaneous* rate of change, or derivatives for short.

Just to reiterate again, no, I am not teaching derivatives to a college algebra or a developmental math course student. If it's not in the standards, it isn't being taught. However, it would be an understatement to say how much I would really like my students, no matter what they are majoring in, to go on and take calculus to see derivatives. So many topics taught in math courses in high school, and in developmental math courses at colleges come back and are tied into topics taught in calculus. Just like *A Tale of Two Cities*, what feels like disjointed topics in these high school classes come full circle to be taught again in calculus.

For example, take the derivative of the function

$$f(x) = \frac{1}{x}$$

using the limit definition. How is this done? First, find $f(x+h) - f(x)$, which would be

$$\frac{1}{x+h} - \frac{1}{x}.$$

Oh, looky here, a subtraction of two fractions using unlike denominators. I teach that in College Algebra! Also, adding and subtracting fractions is taught in developmental math courses such as the Foundations of Mathematics course taught here at Central Georgia Technical College (CGTC), where you need to add expressions such as

$$\frac{1}{2} + \frac{1}{3}.$$

Mathematics builds on itself to tell the ultimate story in calculus of derivatives. To add ½ and ⅓, find the common denominator, which, in this case, is 6; rewrite both fractions using that common denominator; then add. Same thing with the derivative problem, the common denominator is $x(x + h)$, rewrite both fractions using that denominator, then subtract the fractions. If you are curious, the derivative is

$$f'(x) = -\frac{1}{x^2}.$$

Take a calculus class sometime if you want to learn how to do that! However, not only in Calculus I is adding and subtracting fractions coming back, but also look at horizontal asymptotes that are taught in my MATH 1101: Mathematical Modeling course here at CGTC. Mathematical Modeling at CGTC is similar to College Algebra; we just focus a little more on the applications and get to talk about finding the line of best fit for certain data. The topic of horizontal asymptotes in math modeling is brought back in calculus in the form of limits. For example, looking at the function from the derivative example, find the limit as x approaches 0 of it. This happens to be

$$\lim_{x \to \infty} \frac{1}{x} = 0.$$

If you graph the function $f(x) = \frac{1}{x}$ using software such as Desmos, then you would see that the horizontal asymptote of the function, which there is $y = 0$. This is no coincidence to the answer of the limit above being 0, because limits to infinity define a horizontal asymptote.

So students taking these college algebra type classes are missing out on so many other applications of how topics that are taught in high school and other courses taught at the college level are used in further mathematics courses. Would they be interested in knowing about these applications? Sure! If we promote problems based on applications, they would see in their own majors. Could the students at technical colleges be able to handle such classes as Calculus I and Linear Algebra? Not everyone, of course, but why limit someone's education even though many students think of mathematics as a hard and uninteresting subject. Much more needs to be done on how we can unlock the potential of many of our students so that they can bypass the roadblock of "mathematics is terrible" to "mathematics is beautiful."

Before ending this chapter, I want to talk about student work and what we should expect from it at a technical college. As instructors, we should value students' work. There are a few reasons why.

1. There is a less risk of cheating if students are handwriting their work.

This is one I hear a lot from my colleagues, which I slightly disagree with. Just because they are handwriting doesn't necessarily mean they are not using PhotoMath to get the right answer (yes, students, we know about PhotoMath and all the other apps that give step-by-step solutions to problems). They could have the step-by-step solution next to them as they handwrite the solution on a piece of paper.

However, having them show work at all is still better than nothing. It is my opinion that even if a student was using PhotoMath, then by writing or typing the solution, at least they were using resources to get the right answer. I'd personally rather have that than no work at all and cheating, since in the future if they were asked to provide an answer to a question, at least they would have a resource to turn to to get the right answer, but they would also know how to present it in a way to gain full credit. Controversial, I know, but if students, especially in an online environment where it is hard to police, are going to cheat, let's use it to our advantage.

Also, as a side note, I do not think applications that try to prevent cheating work as well. For example, there are multiple software's that prevent students from clicking out of the testing window to open a new tab. However, students can just as easily pull up a phone or tablet and find the solution that way then type the answer into the exam. Really, the only way to police cheating in an online environment is to have students go into a proctored area, but when students have full-time jobs, children, and lives to live, most places are closed by the time they get to the homework/exam. It is not unusual to see students working on homework at 10 at night. So, finding a proctor for each individual student would prove to be difficult. I'm not here to provide an answer to this dilemma, but if work is shown, at least if the student cheated, they know the resources to turn toward in the future to get the right answer.

2. Partial credit can be awarded.

This one might also be a little controversial since some math instructors believe in partial credit, while others do not. At a technical college where students are just wanting the credit to move on to their major courses, I wholeheartedly believe in partial credit. Let's take a look at a student work that was done in MyMath Lab as an example. The question that was being asked was as follows:

> How many points of candy that sells $0.69 per lb ($1.52/kg) must be mixed with candy that sells for $1.32 per lb ($1.32/kg) to obtain 8 lb (3.63 kg) of a mixture that should sell for $0.93 per lb ($2.05/kg)?

The work the student submitted was typed into MyMath Lab, and what this student exactly typed can be seen in Figure 5.1.

After getting the answer, the student submitted to MyMath Lab that for the $0.69-per-lb candy, 4.95 lb were sold and for the $1.32-per-lb candy, 3.05 lb were sold. MyMath Lab marked this as incorrect, since what the student did not say in the shown work was that x represented the $0.69-per-lb bag and that y represented the $1.32-per-lb bag, so the student typed the answer into the wrong box. The student switched the answers around.

$$x+y=8$$
$$.69x+1.32y=7.44$$
$$5.52+0.63y=7.44$$
$$x=y-8$$
$$y=3.04$$
$$x=8-y$$
$$x=4.95$$
$$y=30.4$$

FIGURE 5.1   A presentation of a student's work.

For this quiz, this problem was worth 6 points. For a quiz that is worth 100 points, MyMath Lab counted the answer as incorrect and made it 0 out of 6 points, which is quite a chunk of change in terms of the overall score. Did that student deserve 0 points for this question? It was of my opinion, no. Since the student showed the work and got the right answer, I went back in and actually gave full credit for this answer. Yes, they could have specified what x and y were to be more precise, but why punish a student when they did get the right answer?

One of my biggest goals, and really should be a big goal for mathematics instructors all across the country, should be to build up confidence in their students. Even as a graduate student, I was often not confident at all when submitting homework to the professor. For this student, and many other of my students, mathematics does not come easy to them. So punishing them for getting the right answer is unnecessary. I awarded them partial credit (well, full credit) for a problem they submitted work on and got the correct answer. We can work on being more precise if needed, but it is enough to have the student get the correct answer and move on. This builds confidence in that they *can* do it. They can overcome the problems and get the correct answer.

We do have to be slightly careful with partial credit as instructors, however, if no work is submitted. For someone who has been in the business for a while, I can almost look at a student's answer without work and tell you what they did wrong. That is where I run into trouble, however, with partial credit; I really want to award partial credit because I know what they did. But do I? Actually, not really.

We cannot assume what our students are doing unless we see the work they are providing to us. If the question asked, "What is 2 + 2," and the student responded with "5," one could say "Oh, the student obviously mistyped the answer and meant 4." Well, did they? Just because we think as an instructor 2 + 2 is an easy example, and the mistake on the student's end had to be a mistyped answer, that is not necessarily the case. They might not actually know what 2 + 2 is. Or, when they saw the problem, they might have thought the question was 2 + 3. If we saw the student's work, then we might see "2 + 3 = 5" and know that the student wrote the problem down incorrectly. Then, award partial credit to the student because 2 + 3 is actually 5; great job, student! Why tear them down with 0 points when we can build them up for their successes? Partial credit I am a big believer in, and for technical college math instructors, it should be for you as well.

I also want to note my usage of the word *awarded*. I am a big believer in positive reinforcement, and one of the ways we can build confidence in our students is using positive language such as awarded. Instead of "points were deducted because . . ." words such as "points were awarded because . . ." and then go on to explain what the student did correctly. We, of course, still want to let the student know what they did incorrectly so they learn from their mistakes, but if we start our sentence with "Well, you forgot to distribute to negative, but I will award you half credit because the answer would have been correct otherwise!" rather than "Half points are deducted due to not distributing the negative," we can build that confidence in our students that they are looking for. They did everything right but maybe a few simple errors that can be fixed! Why not prop them up instead of pushing them down?

3. We gain an insight into what our students are thinking.

As an instructor, it is important to constantly be aware of how our students are interpreting and solving problems. If they are showing their work, we can start to gauge what many of them are doing incorrectly if that is the case. Maybe how I taught the subject in class was not a way the students connected with. That happens, and we need to be aware of it. Although many people might think it (just kidding, of course), I am not a perfect instructor. I have good days where I walk into the classroom and leave thinking that lecture was the best I have ever given, and some days I just feel defeated and know the lecture I gave was not a good one. Their work would reflect that day. Or maybe I thought it was a great lesson, but when I see their work, it clearly was not. I might need to reevaluate how I teach that lesson for the future so that the performance of students can improve.

As an instructor who uses MyMath Lab, I can also start to see who follows how I teach lessons and who uses the "View My Example" button. I don't always teach to my students the same exact way the software would show my students how to work the example. A great example of this in my own course is factoring quadratic function of the form $ax^2 + bx + c$, where a is not equal to 1. Take for example this problem:

Factor $9x^2 + 15x + 4$.

There are multiple ways to do this. One method MyMath Lab presents is to use a trial-by-error method (well, that is what I call it anyway). Essentially, we use the fact that we know how to FOIL to guess and check our way to the answer (again, my viewpoint of it). I teach it in a step-by-step methodology because it is my opinion that students like a concrete step-by-step way to get to the answer (plus they can use notes on the quizzes). Multiply a by c to get 9(4) = 36. Find two numbers that multiply to give 36 and add to give 15, just like we did when a was, in fact, 1. Those two numbers would be 3 and 12. Write $15x$ using those two numbers,

$$9x^2 + 12x + 3x + 4.$$

Factor by grouping the first two terms and the last two terms together. From the first two terms, factor out a 3 and an x to get

$$3x(3x + 4).$$

From the last two terms, we cannot factor out anything, so we will write the expression as

$$9x^2 + 12x + 3x + 4 = 3x(3x + 4) + 1(3x + 4).$$

This is fantastic because since there is a $3x + 4$ in both terms, we can factor it out! This results in

$$(3x + 4)(3x + 1).$$

This, I believe, is the best way to factor these sorts of problems. Not everyone needs to agree with me, that is fine. I personally do not teach the guess and check method because I simply do not like it. But students can view an example and if they understand that way better and use it, I am perfectly happy as long as they get the correct answer. However, if I start seeing on their shown work that the majority of students switch to this method (which I do not, the feedback I get from students is that they prefer my method and the results show that is the case), I might want to start considering teaching factoring polynomials in a different manner.

The most important point about student work is that it truly is up to the individual instructor to decide how rigorous a student needs to be when turning in that work. For myself, I am pretty lenient, and if they are getting correct answers in a correct way, I am happy. Some instructors might feel that we should not hand-hold students and expect a level out of them that they may or may not be able to achieve. There are definitely a lot of opinions to be had about the subject of student work, but the number one thing we as instructors need to keep in mind is that we only want the best from and for our students.

Students will rarely fail my class if they are putting in the maximum effort to show work, ask questions, and try their very best in getting the correct answer. If our goal as a technical college is to get them in and out of the door as quick as possible so they can start their careers, that might force our hands to be as lenient as possible in grading student work. It is my opinion that we, however, should strive to build confidence and to build a foundation for students so they can utilize critical thinking skills and problem-solving skills that they can carry onward to other classes and their careers. If we do that, maybe, just maybe, students will leave our classes with more joyful thoughts to mathematics.

The Word *Easy*

Have patience. All things are difficult before they become easy.

—SAADI

A quick Google search of the word *easy* fields this result:

EASY

Adjective

1. Achieved without great effort; presenting few difficulties

2. Free from worries or problems

How easy is it for us to say the word easy while teaching our courses, especially in mathematics? Just the other day I personally got trapped into saying this expletive to my students. The conversation was surrounding finding the slope between two points. It went something like this:

Today, class, we will be finding the slope between two points, which is good news for you all! It is much nicer and easier than those dang application questions in Chapter 1.

What are chapter 1 application questions for the College Algebra class taught at Central Georgia Technical College (CGTC) you may ask? Interest rates, distance equal rate times time problems, that sort of stuff. Any

DOI: 10.1201/9781003287315-6

applications involving linear equations in one variable would fall under this category. For some of you reading this, you would probably *love* to teach these topics; for me, I do not. Oftentimes students will say, "The application problem we did in class looks nothing like the one given in the homework!" or "Word problems are the worst." Chapter 4 of this book goes over how I would answer those questions, so I won't dwell on it too much here.

Going back to the scenario, I hope it is obvious why what I said was wrong. Maybe it is nicer. Maybe it is easier. *For me.*

For me, trying to find the slope between the points $(0,1)$ and $(1,2)$ is "easier" than an application such as

> Gabriella has \$12,000 to spend on her son's education. She will deposit x amount of dollars in a Certificate of Deposit (CD) account at a 3% interest rate, and then deposit the rest in a savings bond account at a 5% interest rate. If the two accounts together generate \$200 of interest, how much money did she put into the CD account?

However, is finding the slope between two points actually any easier than this arbitrarily contrived word problem?

The true answer is who really knows! It is up to the eye of the beholder. It is the same concept as you liking to teach the applications and me liking to teach the more mathy side of the class. Some students understand and thrive when faced with applications. Others look at slope as a foreign language.

As an educator, we need to put ourselves in the desks (or tables) of our students. From that vantage point, there are many questions that may arise.

1. What if a student actually does not find the slope to be easier?

A trained mathematician will have worked with slope for quite a few years. Not only in their high school days of algebra talking about the average rate of change (which is just fancy wording for slope), but throughout their calculus years when talking about the derivative and instantaneous rate of change. Master's- and PhD-level classes would dive into analysis topics, and even some of these topics deal with slope.

To make a long story short, we are trained for this. We have studied it. We know it. For a student at a technical college, chances are they do not. Just because you have studied different mathematical concepts for years does not mean that the student will. This is something I have to keep reminding myself each and every day: as much as I want my students to fall in love with math as much as I have and see the ever-ending beauty it holds, more than likely, students will leave the class and not think twice about that class again.

Also, if a student does not find the slope to be easier, you have lost their trust as an educator. Students trust their instructors to tell them true and factual information. If you tell them it is easier and it is not, then they might spend hours at home wondering why they just aren't getting it. They may agonize over the thought that they really should be getting it, but in reality, what they need is for someone to look them in the eye and be supportive to the fact that not all topics in our math classes are meant to be understood in 5 minutes. And that's okay. Everyone struggles at some point in their life, so struggling with concepts such as slope is okay, too.

Speaking of struggling, one tip I always tell my students struggling with math homework questions is that it is all right to take breaks between problems. I tell them if you spend longer than 20 minutes on a problem, then go on a walk, or work on a different subject. Students who spend hours trying to figure out one problem will not only be discouraged but may give up entirely and fail the class.

2. What if the student does find slope to be an easy topic?

Fantastic! Let's move on then. Well, not so fast. We have to be careful with students that think certain topics may be easy, because in reality, it may lead them down a road of failing.

Let's take as an example Sarah (real name is not given for privacy purposes). Sarah was taking a College Algebra class from me. Sarah would often come to class after a quiz was due discouraged. She would come to me and say, "Professor Youmans, the homework is so easy, but I keep failing the quizzes!" Looking back at the grade book in MyMath Lab, I could see she was telling the truth about the quizzes; she would be getting a passing grade on the homework but a failing grade on the quiz. Since we are speaking on slope, her scores on the chapter about slope were 100, 100, 98.45, 100, 86, and 100 on the homework assignments and then on three attempts at the quiz: 40.13, 7, and 55.

There are several reasons why the student felt that the concept was easy but, in reality, failed the assessment attached to the concept.

A. The Student Didn't Understand Slope After All

This might be the easiest explanation to pass off. Students in MyMath Lab are given two options to help them while doing the homework, a "Help Me Solve This" button and a "View an Example" button. Help me solve this is a way that walks the student through the problem step by step and then gives them credit for the problem if they do each step correctly. This button is a great resource for students who are just starting the homework. Maybe it is the first or second question they are working on, and they are trying to remember the steps that were talked about in class. However, if a student uses the help button for all the questions, are they actually learning the concept? My argument would be no, because at some point the student needs to be able to solve the problem without any support. The quiz has no Help Me Solve This button, and with one quiz being approximately worth 7% of the grade and an individual homework assignment being worth approximately 0.5% of the grade, it's very important they are able to do the homework without support.

The View an Example button is a slightly better tool in my opinion because it gives the same problem worked out but with different numbers. For example, if the student is being asked, "Find the slope between the points (1,2) and (3,4)," the View an Example button will do the same problem but using the points (1,3) and (4,5). Students can use their critical thinking skills to apply what they have learned in the View an Example to the problem at hand to be able to solve it. Again, however, this button is not present when taking the quiz. So, if students get accustomed to solving problems with the View an Example, their environment is changed in the quiz and that would present issues in them passing. It's similar to my (when this chapter was written) 4-month-old. If we start out the night in her crib sleeping, but halfway through the night change her to the bassinet, she doesn't sleep as well due to the change in environment. She has learned to sleep in her crib, not her bassinet. The same concept goes for students. They have learned to use the "Help Me Solve This" or "View an Example" to solve problems, not really taking the time to actually learn how to solve them without the resource. So, when we change the options on the quiz to where they cannot use these buttons, they panic and do not perform well on the quiz.

Another resource a student might turn toward is tutoring centers. Now, before I get into this topic, I do want to say I am a huge proponent of mathematics tutoring. I was a mathematics and statistics tutor while working on my undergraduate degree at the University of Northern Iowa. However, students need to be careful about overutilizing the tutoring center. A job of a tutor is to not teach the material but to guide the student to get the answer on their own. If the student becomes too reliant on the tutor to get the answer, when it is time to take an assessment, they will not be prepared.

I highly encourage all math faculty to work closely with their tutoring centers. For example, when I was hired at CGTC, I went straight to the Academic Success Center (ASC) coordinator and talked with her regarding how to help our students be more successful in technical college math courses. Tutoring can be a wonderful resource for students, and working with the center can help mitigate the issues of students becoming too reliant on the tutors to solve problems. Students will also be able to see that the instructor wants nothing but the best from them if they understand that the instructor is pushing tutoring.

B. Test Anxiety

Much research has been done on test anxiety, so I don't want to spend too much time on this subject. Though, I would be amidst if test anxiety was left off the reasons why students do not perform well on quizzes and exams. According to Rizwan Rana and Nasir Mahmood in the article *The Relationship between Test Anxiety and Academic Achievement,* they state that "test and examination stress is thought to prevent some individuals from reaching their academic potential." They go on to say that "pressure of scoring high on tests, fear of passing a course, consequences of failing in test and incompatibility of preparation for test and demand of test were the reason for cognitive test anxiety."[1]

These statements seem to back up my statements in part A, that a lot of test anxiety comes from the use of different resources such as the Help Me Solve This and View an Example buttons. By exposing these buttons to them, they might not be ready to take the quiz and therefore do not perform well on it. This does not necessarily mean not giving this as an option to students if you use MyMath Lab or other software, but it is important to communicate with students the importance of learning the concepts on their own so that they are able to perform well on different assessments given in the class.

It is quite disappointing when students do not achieve their full potential when faced with test anxiety. To help mitigate this issue here at CGTC, we give students two attempts on each quiz. I specifically tell my students to try to take the first one, and if they do not get the grade they want, utilize the ASC or my office hours to discuss what went right and wrong on that quiz. We can help them prepare for that second attempt. We also let them take the quiz at home in a comfortable environment. Again, environment plays a big role in how students perform, and if they are in a comfortable environment, students tend to perform better.

However, as the article suggests, another reason a student might develop test anxiety is due to the demand of the test. We need to be careful when writing quizzes and exams. We do not want to create an exam that will be too demanding on students, whether that be time demand or the exam being too difficult. In terms of time, advice that was given to me was to take the exam yourself while timed, then multiply the time it took you to complete it by three and that is how much time you should give students. That advice has worked well for me in the past. In terms of an exam being too difficult, I would say that it takes experience to learn how to write a good exam. It's a good idea though to let a colleague take the exam and give advice on it. While they also have a background in mathematics, they are still someone new looking at it and can give you pointers on how a question is worded, the length of the exam, and any other advice that person might give.

Many of the students taking math courses at CGTC are in a health-related field. These health-related fields are very selective when it comes to accepting students into the program; many students need As and Bs to even be considered for selection. So, the pressure to perform well on quizzes is exacerbated by the fact that students know they need to perform well to be selected into the field they want to have a career in. On top of mathematics being an already feared course for many of these students, the fact that more than a C might be needed for their grade compounds into having text anxiety.

To help combat this issue for my students, I first have them remember to utilize their resources, including me as their instructor. My sole job is to help them succeed in the course and to remind them of that helps them stay confident in their abilities to perform well. Also, giving them two chances on the quizzes instead of one has helped since they know if the attempt goes badly, they have another opportunity. Of course, on that second opportunity, test anxiety might increase even more since they would know it is their last chance to perform well. Having students submit work will help as well, so the instructor can award partial credit. Finally, if the student performs well, it is important to recognize them for that. Walking into class saying, "Hey, Mark, great job on that quiz!" does wonders for their confidence in the course. Congratulate your students on their successes, help them during their failures, and test anxiety in your students will be a smaller factor.

C. Technology

Technical colleges have a high population of nontraditional students, which includes people in older generations. First of all, what great joy it is teaching this type of student. Many of them have full-time jobs and children to take care of, so for them, coming back to further their education is a huge step in their lives. Sarah was a nontraditional student. Being that homework and quizzes were given through MyMath Lab, Sarah, and many nontraditional students, have a hard time doing math homework in an online environment. As stated earlier about being in a comfortable environment, Sarah was not while doing assignments and assessments online.

Now, I don't want to say all nontraditional students have trouble with technology, just like how not all Generation Z students are comfortable using computers and technology. In fact, some students who are older than me in my classes are way better at technology than I am. However, it is true that for many nontraditional students, when they went to school, most work was done with paper and pencil. So, by having them work in MyMath Lab and trying to learn the software, they perform negatively because not only are they trying to learn the mathematics but they are also learning how to input their answers into a computer software. They may have performed better if the entire class was based on paper-and-pencil assignments.

Let me use another student of mine as an example: Linda (again, name is changed for privacy). Linda was a nontraditional student taking a College Algebra class with me online. She was wanting to start a career in nursing, which was a complete 180 from the job she was doing while going to school. Linda performed badly on her first quiz, so she sent me an email stating that MyMath Lab was getting in the way of her performing well. As stated earlier, it was important for her to get an A in the class due to selection into the nursing program.

Instead of having Linda do the quizzes online, I printed to PDF the quizzes and sent them to her in an email so she could print it out and send it back to me. I only did the quizzes and the midterm and final since the homework had unlimited attempts on them. She often attended online office hours I provided to the class through Webex to discuss questions on the homework, whether that be about the technology or the math in general. Through working hard and the fact that she could do pencil-and-paper for the assessments (which the quizzes and exams combined was 80% of the grade), she passed the course with an A and was selected into the nursing program at CGTC that next semester.

Yes, it took longer for me to grade her assignments than others since for the others, MyMath Lab automatically grades the assignments. For hers, I had to hand-grade them. However, if our goal is to provide our students with the best possible chance of success, spending an extra 5 minutes grading an assessment was no problem for me. This is compounded by the fact that MyMath Lab, when printing assignments to PDF, will print out the answers as well, so I did not manually have to do the assignments before grading them. I did give this option to the entire class that specific semester for fairness to other students who also would be struggling with the technology. As long as they have bought the software, they can print out the assessments and turn them in to me.

For me as a younger instructor, it is perceived as an easy task for students to utilize the MyMath Lab software. However, I often get complaints of this software on the difficulties of using it. So, I need to keep in mind as the instructor that for students to be successful, not only do they need to know the material, but they also need to know how to use the software to give the answer. Maybe that means taking a day out of instructor to teach on how to use the software. Or maybe I should emphasis more often to attend office hours if a student is struggling with technology. Whatever the case may be, my sole focus is for a student to learn the material, not necessarily learn how to utilize a new technology.

Also, I need to realize that oftentimes, software such as MyMath Lab is not perfect in giving the correct answer to students depending on the directions it gives. For example, if the software says to round to the nearest hundredth and the answer is 3.17 but the student inputs 3.2, the software will mark it as incorrect, although they technically did give the right answer, not just in the form they wanted. Does a student deserve a 0 out of 6 points on this problem due to not rounding correctly? It is my belief that no, they should not. In fact, for many students who are taking mathematics at a technical college whom math is a struggle for, I award full credit. While they did not follow directions, they did get the correct answer, and thus, I believe they should get full credit. If nothing else, a very slight deduction in points (maybe 0.5 points) and in the comments saying to remember to follow the directions. Why throw the book at them when the error in the answer the student gave was small?

Going back to Sarah, long story short she decided to withdraw from the course and try to take it again in a future semester. That in and of itself is a brave decision for a student, one that the student should be proud about. Unfortunately, no matter how hard it seemed I tried to help, she was not understanding the material and failed three quizzes before withdrawing. Withdrawing does come with its own negative consequences, for example still having to pay for the course up to the point of withdrawing. However, taking a W does not affect her GPA while taking an F would (and you still have to pay for the course). Recognizing that you are not quite ready for a college-level mathematics course is important so that she can retry either in a further semester or take one of our learning support classes while taking College Algebra. She may also want to consider taking a developmental math course before taking College Algebra. It might also be the case that I was not the correct instructor for her learning needs. That happens sometimes, and it is important to recognize that not all students will like the way you teach. So it may be discussed to take the course over again taught by someone different. All these options can be discussed with me or her academic advisor at the college so that she can be placed in a position to succeed.

This brings me to another point about Sarah, and many of our other students in technical colleges, and that is, are our students actually ready to take a college-level mathematics course? Again, as I, the instructor, perceive this college-level course to be an "easy" course that everyone should pass, this is clearly not the case when only about two-thirds of the students at CGTC are passing College Algebra. So, what happened to the other one-thirds? Well, the earlier facts about test anxiety, technology, and not understanding the material still stand when it comes to failing the course. Out of not understanding the course material, should that student have been placed in College Algebra in the first place?

With the rise of COVID-19, CGTC and many other colleges around the country suspended testing requirements. This includes mathematics placing testing. Before, students would be placed into courses depending on what they scored on the exam. Now, students, no matter how long they have been out of school, are taking College Algebra whether they have the prerequisites or not. As of this writing, testing requirements are not coming back either.

If students do not have the prerequisite knowledge for College Algebra, of course they are not going to perform well. If I was asked to bake a cake but was given no recipe and was asked to complete the cake based off one

50-minute demonstration, I would probably not perform too well, especially since I have never baked a cake from scratch before! So why should we expect students to perform well given no background knowledge of the subject?

To back up my point, the numbers for some of our mathematics courses with testing requirements and without testing requirements are stark. Let's take a look at the fall 2020 numbers with testing with the fall 2021 numbers without testing (Figure 6.1). The attainment rate is students who received an A, B, or C in the course divided by the total amount of students taking that course. Withdrawals are not counted. College Algebra, our most populated course at CGTC, is MATH 1111.

The only positive gain was MATH 098, a developmental class. All other classes have a decrease in students passing the course after testing requirements were discontinued, the highest being our Math Modeling course,

TABLE 6.1   Comparison of attainment rates for fall 2019 and fall 2020 for select math courses at CGTC.

	Fall 2019 (with testing)	Fall 2020 (without testing)	Percent Change
MATH 0098	63%	67%	6%
MATH 0099	70%	67%	−4%
MATH 1012	62%	59%	−5%
MATH 1101	77%	58%	−25%
MATH 1103	86%	69%	−20%
MATH 1111	72%	67%	−7%
MATH 1113	78%	75%	−4%
MATH 1127	58%	49%	−16%

TABLE 6.2   Comparison of attainment rates for spring 2020 and spring 2021 for select math courses at CGTC.

	Spring 2020 (with testing)	Spring 2021 (without testing)	Percent Change
MATH 0098	85%	79%	−7%
MATH 0099	73%	63%	−14%
MATH 1012	71%	62%	−13%
MATH 1101	90%	78%	−13%
MATH 1103	100%	73%	−27%
MATH 1111	74%	61%	−18%
MATH 1113	95%	79%	−17%
MATH 1127	86%	77%	−10%

which is often taken by students who are ready to begin courses in their programs. This trend continued onto to the spring as well (Figure 6.2).

This did not just happen for mathematics courses in the general education department at CGTC. ENGL 1101: Composition and Rhetoric is another popular course for a lot of our students who need to take general education courses as part of their diploma. From the fall of 2019 to the fall of 2020, the number of students passing English fell from 73% to 66%, a percent change of −10%, and in the spring of 2020 to the spring of 2021, it fell from 77% to 64%, a percent decrease of 17%.

The dean, the division heads, and the program chairs (including myself) met to discuss these numbers and to see what changes can be implemented to improve attainment numbers. Two changes were immediately discussed and implemented to start in the following semester:

1. Place more emphasis on the diagnostic exams.

2. Increase the awareness of the ASC to the student.

Regarding the first point, many of our general education courses have a diagnostic exam that was put in place with testing being removed. For our College Algebra course, this diagnostic is 10 questions that students should know how to do before taking the college-level algebra course. For the couple years it has been in place, faculty members have not been using it to its full advantage. Students would just take the diagnostic, get whatever grade they got on it (the grade does not count toward their overall average), and move on without paying attention to the score. Now, we are insisting faculty members to place more emphasis on the diagnostic exam to help students realize that they might not be prepared for the course. We are giving students the following information to help them diagnose the score:

90–100: No additional support recommended. Student should expect to periodically require assistance on a few questions in each homework set. Student should be prepared to spend 3–4 hours per week to complete assignments.

60–80: Recommend 1–2 hours of individual tutoring per week. Please see the tutoring options listed in the Course Information page of this course. Student should expect to encounter more difficulty in completing many homework sets. Student should be prepared to spend at least one hour per day on this coursework (7–8 hours per week) to complete assignments.

50 or below: Recommend 2–4 hours of individual tutoring per week and to register for MATH 99: Learning Support. Please see the tutoring options listed in the Course Information page of this course. Student should expect to encounter difficulty in completing most homework sets. Student should be prepared to spend at least 10 hours per week to complete assignments.

We hope this information will help students realize where they are in the course, and if they feel like they are not prepared, either withdraw the course and take a lower-level course first, take a learning support class, and/or utilize the ASC.

Speaking of the ASC, our second recommendation to faculty was to make sure students know about and use the ASC. The ASC helps with more than just mathematics. For instance, for an English course they can review papers and make suggestions on grammar and spelling. One recommendation for faculty was to take 20 minutes out of class and walk the students down to the ASC. That way, students knew exactly where the ASC was located and could meet a few tutors that would be able to help them. We also encouraged faculty if they did not want to walk down with their students to take time out of their class to have representatives from the ASC talk with the class about their services. That way, the ASC would come to the class, not the other way around. I personally created an extra credit assignment for my students which involved them getting a signature from a tutor stating they visited and chatted with the ASC. It was worth 10 extra-credit points on a quiz.

As of this writing, no data has been given for attainment numbers that can be used to compare the numbers before the two recommendations of utilizing the ASC and placing a heavier emphasis on the diagnostic exams. We obviously hope numbers will increase to where testing was required, and even if not, we will reconvene and see how we can change our courses to promote student success.

Overall, the word *easy* or any synonym of the word should be outlawed from our vocabulary while teaching. Our students come from different backgrounds and are studying different subjects, and at technical colleges, mathematics is usually not a subject students want to major in if they want to transfer to a 4-year university. And that is okay. However, if we expect the mathematics in our classes to be "easy" for them, that would be the first mistake in creating a positive classroom environment.

Take, for example, a developmental mathematics course where part of the standards is to add and subtract whole numbers. One example of this type of course taught at CGTC is named Foundations of Mathematics. As someone with a master's degree in mathematics, adding and subtracting whole numbers is second nature. So should I expect that same thought from students taking the course? Maybe, maybe not, but it is always better to be on the side of caution, especially when only 62% of students passed the foundations class in the spring of 2021 (the class labeled MATH 1012 in Figure 6.2). So, for my online course, I had in the summer of 2021, I still created an online video that students could watch when adding and subtracting whole numbers. I did not mention the word *easy* because doing so would risk a student not thinking the subject was easy. I still held online office hours during the time they would be doing the lesson, because students could still have questions on it. The result was that the averages for the lessons on adding whole numbers and subtracting whole numbers for the summer of 2021 were 98.25% and 97.5%, respectively. The quiz on operations on whole numbers resulted in an average of 97.38% (disregarding one zero for the student not turning in the quiz).

I treated the lessons on adding and subtracting whole numbers exactly how I would if I was teaching someone how to take a derivative in Calculus I. For students who have been out of school for 10 years, it wouldn't matter if we are adding or subtracting whole numbers or taking a derivative. Both topics deserve the same amount of respect and time given to each lesson.

It is important to remember that we as instructors were once in our students' shoes. We once learned about adding and subtracting whole numbers just like we once learned how to take a derivative. You can make a difference in students' lives just by telling them it is all right not to understand a topic, and then spending time with them to understand that topic. While teachers in their past may have sped by or not given enough attention to the topics they think are easy, let's pay special care to how we perceive and then teach mathematics to our students. Let's not assume any concept in mathematics is "easy." If we do this, I'm betting that students will try their best to get a passing grade in your course, which would mean they give you their 100%. When students try their best and are productive in the class, that will in turn make our jobs as instructors "easy."

REFERENCE

1. Rana, Rizwan and Mahmood, Nasir. (2010, Dec.). The Relationship between Test Anxiety and Academic Achievement (December 2, 2010). *Bulletin of Education and Research*, 32(2), 63–74. Available at SSRN: https://ssrn.com/abstract=2362291

CHAPTER 7

A Personal Cheerleader

N O CLICHÉ QUOTE AT the beginning of this chapter, you ask? Sorry, friends, I just could not find the perfect quote to fit this chapter. Is there one out there? Probably. What I want to say in this chapter, however, just doesn't live up to any quotes I can find out there. What I say in this chapter may seem obvious, but I think it is important to remind ourselves day in and day out why we do what we do, and that is that we love to watch students succeed. Watching their successes and knowing you made an impact on their life creates a sense of fulfillment for oneself. *You make a difference in students' lives, whether you see it or not.*

So, let this chapter be a motivation for you and your students. Read all the motivation quotes you want, but at the end of the day, it is up to you to make your students feel like you care about them and their successes. In Chapter 3, we discussed how getting to know your students creates a huge advantage to their successes in and out of the classroom. There is something else, however, that I tell my students that I think makes a huge impact on them and one that each and every student needs to know: that no matter who supports you or not, you always have one supporter (cheerleader, if you will), that being me, your instructor.

That's right! I am your own personal cheerleader. No, I am not standing out in front of your window with pom-poms and singing fight songs, but I am staying awake at night hoping you will pass quiz 2 of the course. I am at my desk grading, hoping that you have shown improvement from the midterm. I'm thinking about what you are doing now, a semester, a year, a few years after you took my class. My job as a mathematics instructor at a technical college is to get you to graduate. That means to pass my class.

DOI: 10.1201/9781003287315-7

That is not all I want, however; I want you to learn the skills necessary to succeed in college and in "the real world," as they say. I want you to learn a thing or two about why I personally love mathematics and love teaching it. I want you to know the resources you can turn toward if when you leave my class things go south. *I want you to succeed.*

It is my opinion that students do not hear this enough. For example, many students spend a lot of time doing homework and quizzes through MyMath Lab, which could stand in the way of the students' success. Just looking at one of my own online classes being taught in the fall of 2021 (College Algebra), one student so far has spent 53 hours on assignments, while another one has spent 102 hours. Quizzes in my class have a time limit of 2 hours, and more often than not, students will take the full 2 hours. Students spend a lot of time working problems, reviewing lecture videos or notes, and taking assessments. So, when they get problems incorrect and it took them a long time to even get an answer, it might bring them down. Whether it is figuring out how to solve linear equations in one variable or factoring a quadratic polynomial, students seem to get discouraged and give up easily. Why do they do this? Well, for one, the demographics of students at technical colleges show that many of them have children, have full-time jobs, and are taking classes at the school to either advance in their career or change it completely. A lot of students do not have time to do the work.

Another factor that may seem impossible is the fact that the student may feel as if they have no support system. For one student, they may have no family or friends to help guide them on the path to college. For another, they may be struggling financially, finding it hard to put food on the table. At technical colleges (and really any higher education school), these are the types of students you see more often than one might think. We as instructors need to remember this about our students: they are not perfect. They have lives outside of our classroom. Yes, we do want to teach them time management skills, but what happens if the night that work was due the power was cut off due to not being able to pay the bill? Will a student disclose to you that information? Maybe, maybe not. But in the grand scheme of what we are trying to do at technical college, does turning in a mathematics assignment late by a few days really matter? For student fairness to those who did turn in the assignment in on time, maybe. However, for a student who is not only struggling in school but in life, giving them even an extra day past the deadline might be the difference between them graduating or not. That may seem extreme, but there are a countless number of times that students have given up after even one homework assignment or quiz. Let's look at some withdraw numbers from Central Georgia Technical College (CGTC) to show this.

At CGTC, students have three days to add or drop a class without a W being placed on the transcript. After that and into the 10th week of class, students can withdraw from the course with a W. After the 10th week, it is an automatic F (unless during COVID-19 when students could end the semester with a Z grade, which produces a similar effect to taking a W). The numbers that are about to be provided are only for those who withdrew from the course with a W or Z grade.

At CGTC, 804 students in the spring of 2021 took College Algebra. Of those 804, 111 of them withdrew from the course with a W or a Z, which is 14% of the students. That is quite a bit! A possible explanation for this number is that students tended to give up after a few assignments, especially due to the nature of COVID-19 in the spring of 2021. Getting the numbers for how many did only one or two assignments before withdrawing would be nearly impossible, but it would be a reasonable guess to say at least half of the 111 had only done one or two assignments through the entire semester.

Another popular class at CGTC is MATH 1012: Foundations of Mathematics. This is a developmental math course that starts with whole numbers and what to do with them and then builds up the students for some applications in geometry, measuring, and statistics. Out of the 329 students which took the course, only 22 withdrew with a W or a Z, making the percentage of withdraw 7%. This is not as high of a number as College Algebra, and again one could hypothesis why this is true. The content is perceived easier, as you start with whole numbers and not linear functions. Students taking the class are more motivated to take the course due to it being a required course for graduation (although many times College Algebra is as well). Whatever the case may be, it is important as a department to look at the numbers to try to fix them so that more students can succeed in the course.

One more class as a case study is our MATH 1127: Introduction to Statistics class. This class is usually not a class taken by the general audience at CGTC. Students in these classes usually are there for a specific reason, usually a department-specific reason. Departments that students can take statistics from are associate's degree programs such as general studies, engineering technology, metrology, and business management. One hundred twenty-six students took Introduction to Statistics in the spring of 2021. Of the 126, only 7 took a W or a Z, which is a 6% rate. Students are more motivated to take this course, because again, it is used for department-specific requirements and not general education requirements. Students are less likely to give up after one failed homework assignment or quiz. Those taking College Algebra, on the other hand, which students across multiple disciplines take for the general education requirement, are less motivated to take the course due to the nature of not understanding *why* they are in the class. When you do not understand the why, then failing assignments provide a lack of motivation to try, and you start to see the higher D, F, and W rates.

If we show kindness and fairness to our students, they will, in turn, give us 110% on their assignments. If we show that we truly and deeply care for and love them, they will show that right back and maximize their efforts in your class. Creating an environment, especially in a mathematics classroom, where mistakes can happen and we learn from each other is a great way for the classroom to be a safe area. We are cheerleaders for our students, and if on a daily basis we show our students that we deeply care about their learning in and out of the classroom, they will feel supported.

How does this look like in the classroom? Let me give some tips on how to show your students that you want them to succeed.

1. Tell them!

"Well duh!" you say. Seriously though, tell them. Make a purpose to tell them that you care. Make a purpose to tell them that you want to succeed. Just telling them is not enough. Make a purpose to be in class 5 to 10 minutes early to talk with your students about their week and their lives. If they see that you care about their personal lives, then they will be inspired to work harder and succeed at what they do. Make a purpose to stay after class 5 to 10 minutes to let them ask questions. If you do not allow questions to be asked, how will they learn the material? Tell them you want them to succeed, then back it up by action.

One comment I try to make a point of telling my students is that I do not expect them to learn all the material we covered in an hour and 10 minutes during that time frame. I did not get a master's degree by walking into class and then walking out thinking I could take the test the next day. Mathematics, and really any subject, requires dedication and time to learn. While students at technical colleges are keen on taking classes in nursing, engineering, computer science, and so on, part of the mission of any degree-level college is to create well-rounded citizens. Apart from that goal is to have them take classes in mathematics, English, and humanities. So getting your students to understand why they are taking the classes they are taking will go a long way to them spending time learning and studying the subject they are taking a class in.

2. Work examples with a purpose.

Each example done in class needs to be done with a purpose. A lot of times the purpose is to demonstrate a problem they will see in the homework. However, why does that always have to be the purpose? Maybe one example might be done to explicitly see where students might make a mistake on. A great example of this is when students start talking about solving linear inequalities in one variable in a College Algebra class. An example could be as such:

Solve the linear inequality and state the answer in interval notation.

$$-2x + 1 \geq 3$$

This example is an example of something they will see on the homework, but the purpose of doing this example will not be to get the right answer. In fact, for the instructor, the purpose will be to get the wrong answer. Let's look at the steps to do this. We need to isolate x; to do this, we subtract 1 from each side which results in

$$-2x \geq 2.$$

Dividing by −2 results in

$$x \geq -1.$$

In interval notation, the answer is therefore

$$\left[-1, \infty\right).$$

When I tell my students, "And this is the right answer!" nine out of ten students in the class shake their heads up and down, because let's be honest, most students will not question what the instructor does. But when I look into the students' eyes and tell them the answer is, in fact, incorrect, many students give quizzical looks. This is a true teaching moment, one that can extend further than mathematics. We all make mistakes. In this one, one of the rules of inequalities is that when you divide or multiply by a negative number, the inequality is flipped. Meaning when I divided by −2, I should have actually written

$$x \leq -1,$$

and therefore, the answer in interval notation is

$$(-\infty, -1].$$

Making mistakes like these shows the students that you care about their learning. You want them to succeed in solving linear inequalities, so you show them how one could make a mistake and then how to fix it. This is a great way to grow community in the classroom where mistakes are welcome. Speaking of mistakes . . .

3. Do not pass up an opportunity to learn.

We learn by making mistakes. In mathematics, mistakes happen all the time, even when I am doing my own work (believe it or not, I am not perfect!). So, while you may create examples that have intentional mistakes, you might also make mistakes on accident. *Use this as an opportunity to learn with your students.* Do not let these accidents be of waste. Have your students see why you made the mistake you did. Show them how you knew there was a mistake at all. A great example of this is something I did in my class the other day, working an application problem about interest. Going back to Chapter 6 for a moment, the application was

> Gabriella has $12,000 to spend on her son's education. She will deposit x amount of dollars in a Certificate of Deposit (CD) account at a 3% interest rate, and then deposit the rest in a savings bond account at a 5% interest rate. If the two accounts together generate $200 of interest, how much money did she put into the CD account?

Without going into detail on how to solve such a question, when I worked the problem for my students, I got the answer of −$8000, or something like that. Immediately, I knew the answer was incorrect. Heck, I knew it was incorrect a step or two before getting to that answer. But I finished it out anyway, why? *Because this is a learning opportunity.* Why does the answer of −$8000 not make sense? For one, does a negative answer to this kind of question make sense? Not really, because that would suggest the bank is paying her money to invest in these accounts. If you know a bank that does that, please point them in my direction! Where did I make my mistake, you ask? I wrote 3% as 0.3 instead of 0.03. The students see the mistake and know that even a man with a master's degree in mathematics can do something as simple as writing 0.03 and 0.3.

This mistake may seem pretty inexcusable. However, for a student who has not taken a mathematics class in 20 years? This mistake is a little more common. So, I spent a few minutes, not long, showing them why my answer was wrong and how I knew it was wrong, because if (and when) they make the same mistake, they now know how to handle it.

Not only do the students see the mistake and hopefully learn from it, but they also see you as more human. As a mathematics tutor at the University of Northern Iowa, students would often not speak to the instructor about issues they were having in the course. The number one answer to the question about why they have not asked their instructor is that they were scared to approach the instructor. First off, that answer has always disappointed me. As a mathematician, I love to talk about math. I will talk for hours with students or other faculty members about different topics in the math universe. So, when students come and want to ask questions about their homework, I relish that opportunity. Not only that, but I also get to know the student better. I get to ask them what their major is. I get to learn a little bit more about how the student thinks, and what their thought processes are while working through problems. This helps me in my lecture, so I can start to style the lecture to the students' needs, not just what I feel is the best. Now, I am not saying to teach in an uncomfortable way. However, if I can slightly change how I teach or even spend more time on a topic because I know students are feeling lost or confused, that goes a long way to student success. Let me give an example.

One lesson in a College Algebra class we teach is on applications of linear equations in one variable. An example of such a problem is Gabriella's son's education a few paragraphs ago. My first semester teaching the section I took examples from the book, and I presented them in class. Many students would stay after class to ask questions about that lesson due to the nature of applications: every problem is different; thus, students have a difficult time transitioning from one problem to the next. So, instead of presenting from the book, I started presenting the questions directly from the assignment knowing that the questions the students get on the homework will be similar but with different numbers inputted into the problem. I made this change during the fall 2021 semester. Just as a comparison, the spring 2021 average on that homework assignment (disregarding the zeros for people who did not do the assignment) was 75% for eight students. Then, in the fall of 2021, one section of 12 students who completed the assignment had an average of 79.5%. Another section of 13 students had an average of 81.7%, and the last face-to-face section I taught in the fall of 2021 had nine students complete the assignment with an average of 77.73%. The rate of change for those three assignments are 5.5%, 8.5%, and 3.2%, respectively. Not a hefty change, but it is still a slight positive change that could have more than a slight effect. This lesson is the second lesson in the course, so giving an assignment that is perceived as difficult could create a sense of dread for the rest of the semester. Many students at our college also have full-time jobs and children, so if it takes 5 hours to complete a homework assignment, they tend to give up more easily. The statistics show that even making small changes could have a positive effect on the performance of the class.

So, please do not be the instructor who erases mistakes and puts them off as if they were just quizzing the student. If that was your intention, fine. But if you made an accidental mistake, own up to it. Apologize if needed. Show that you are a human being who knows that one mistake will not end the world. We learn, we grow, and we show that we care about how our students learn.

Let me give at least one example of this in action where I unintentionally made a mistake. I was working an example of rational inequalities in MyMath Lab, and the question asked students to solve the rational inequality

$$\frac{10}{x-2} \geq 2.$$

The first step I tell students to do is to make the rational inequality "set equal to 0," meaning to get 0 on one side of the inequality. To do this, subtract 2 to both sides, which results in

$$\frac{10}{x-2} - 2 \geq 0.$$

Now, subtract the rational expressions by finding a common denominator to get

$$\frac{10}{x-2} - \frac{2(x-2)}{x-2} \geq 0.$$

Simplifying results in

$$\frac{10-2x-4}{x-2} \geq 0,$$

and simplifying further results in

$$\frac{6-2x}{x-2} \geq 0.$$

To make a long problem short, I got the answer to be, in interval notation,

$$(2,3].$$

I showed them how to input the answer into MyMath Lab, and what did MyMath Lab tell me? Wrong, try again!

"What?" I asked myself, "I'm surprised that was wrong; let me take a look at what I did incorrectly." So, for the next 5 minutes, I stared hard at the problem, trying to piece together why the answer was incorrect. Maybe by now you, the reader, knows what the issue is. I could not figure it out. Then, it dawned on me.

$$10-2(x-2)=10-2x+4$$

Not -4 as I originally suggested. I incorrectly distributed. A negative times a negative is a positive. Six years of college work, a master's degree, and multiple years of teaching I should have known that. But I am human. Not to try to make excuses for myself, but at the time I did this problem, I had a 4-month-old at home. I didn't get much sleep the night before. I didn't tell any of that to my students, because I did not want them modeling the behavior of trying to make excuses for every mistake we make in life. I owned up to it to my class. "Sorry, guys and gals," I said, "that is my fault. I apologize." After apologizing, I showed them why the error was, in fact, an error and got the correct answer when I made the fix. That showed them that hey, this math instructor dude might be crazy, but at least he is human and can and will make the same mistakes I do occasionally.

So do not let mistakes pass without having your students learn from them. We all will make a mistake or two (or a lot, in my case) in our lifetimes, so let's teach our students how to handle that adversary, correct it if it needs correcting, and push through to succeed at our given task.

4. Online classes are more than just self-guided.

At technical colleges, since there is a very diverse group of students, many classes are offered online to accommodate those who work during the day, for those who live outside the service area of the college, or for any other number of reasons students do not have time or motivation to take in-person classes. Just to give you some numbers, 5,493 students took a non-face-to-face course, such as an online, hybrid, or enhanced course at CGTC. The face-to-face population was 4,205.

In the general education department at CGTC, online courses are a big topic of discussion that we are trying to have more buy-in from our faculty. Pandemic or not, higher education is leaning toward online education, especially with the advancement of educational technology.

Especially in an online environment, students still want to feel like they are cared for and supported by their instructor. There are a few ways that an instructor could make this happen.

A. Create weekly announcements.

Creating weekly announcements helps students know that you are still right alongside them, watching them progress in the course. Normally, I will put whatever is due that week (if anything at all) and then tell them to have a great week and remind them to ask questions if needed. My number one goal with this is to make sure they know that they are not alone when it comes to the online course.

Oftentimes, and this happened when I was taking my own online course, students forget that the instructor is there to help them. So, they think everything they have to do is by themselves. Not true at all. By reminding them to ask questions, I am reminding them that I am here as a resource for them. It mitigates the risk of cheating, because they don't have to look up all the answers to understand what is going on. I am right there, cheering them on to success and helping them when they are needing help.

B. Hold online office hours.

This one will also help students not feel alone in the course. CGTC uses Blackboard as its learning management system (LMS). I have also used Canvas as well, but others may have used Google classroom or Moodle as other examples. We also use Webex as our video system, but other colleges might use Zoom. In Blackboard, I create a specific announcement and title it "Online Office Hours." I post the link so that when they click on it, it takes them straight to a videoconference where I will be waiting and willing to help with whatever questions they need answered.

It is important to be on time if you are thinking about holding online office hours. What we don't want to have happen is students waiting for their professor to join the conference. Just as an instructor's time is precious, so is the students'. Also, take the opportunity to get to know the student. Ask them questions about what they are majoring in and what their specific dreams are after college. This develops a rapport with the student, and they can leave to conference knowing that you care about their success.

C. Create your own videos/notes.

Students often go the whole semester without even hearing their instructor's voice as they move along in their online courses. We as instructors can help mitigate this issue by creating our own videos and maybe even letting students have access to notes you use as you create those videos.

I also like to selfishly think the way I teach things is the best way to do certain problems. Sometimes, MyMath Lab does not expand on an issue as I would like, or the Khan Academy or YouTube video is just not up to par with what I would like to see. So by creating my own videos, I can control the narrative and what gets taught. Students can gather the information that is most important, rather than trying to parse YouTube or Khan Academy for the information they are needing to solve a problem.

D. Create your LMS course with care.

When your course goes live on the first day of class, you don't want a student logging into your LMS and seeing a course that is disorganized and sloppily put together. Make sure before your class is published to use the student view if one is provided. This lets you see exactly what the student will see, and it helps you organize the class in an efficient manner. Have another faculty member browse your course as well and have them offer up any tips that will provide the students with the best chance of succeeding in your course as possible.

Online courses are the way of the future. If we put a little bit of love and care into them, we can try to simulate what it would be like if the course was taken in the classroom. Online students deserve as much of a quality education as a person taking the course in person. By having well-cared-for online courses, we can guarantee online students can feel they too are getting a well-rounded education.

5. Give your students an opportunity to provide feedback.

One of my favorite ideas I learned in graduate school when I was beginning teaching was to use exit tickets. According to Larry Wakeford at Brown University, an exit ticket can "provide feedback to the teacher about class, require the student to do some synthesis of the day's content, [or] challenge the student with a question requiring some application of what was learned in the lesson."[1] When you give students an opportunity to provide feedback, they will see that not only that you care about their learning but that you as an instructor care about improving oneself as well.

I preach a lot in my classes the idea of improvement; as long as you, the student, are showing improvement as the semester goes along, that will go a long way in helping you succeed in the course. However, I, as an instructor, should always try to be improving my own teaching as well. I am not a perfect instructor, no matter if I have 1 year of teaching or 30. I want to model the behavior of improvement to my students, and a great way to do that is exit tickets.

The way I do exit tickets is this: on handwritten exams, I provide sometimes an optional or sometimes a graded question asking for the student to reflect on the class so far. Of course, you get a lot more responses if the question is graded, so I would suggest making the question worth points. Some questions you could ask are, How are you doing this semester? What is going well so far? How can the class or you improve? As an example of these types of questions in work, most recently I gave out a midterm in my College Algebra class. The optional question I asked was this:

Let me know how the semester so far has been going for you, whether it is in this class or other classes. What has been going well so far? What do you think can be improved upon?

Out of 34 students who took the exam, 16 students answered the optional question for a rate of 47%. Here are some examples of the responses I received. I did not change any wording to make them grammatically correct.

- So far the semester is going, I'm enjoying the class & the work load is not hard & very understanding. Something I can improve in is my attendance!

- It has been alright, I need to improve on my time management.

- Doing terrible got a lot going on.

- Shit, but I'm fixing it.

- This is the only class I am taking. This is the furthest I have made it through a school year.

First off, notice just how different each response is. Some are positive; some are negative. As an instructor, it is important to realize that students have a lot going on in their lives outside of the classroom. So, the feedback I give these students on their responses is critical. For the first bullet point, not much feedback is needed. They say they are doing good, they got a decent grade on the midterm, and the overall grade in the class is passing. But going to the second point, this student did fantastic on the midterm, but at the time of this writing, had a failing grade in the course from missing five assignments. Student C says they are doing terrible, but in fact, after calculating their grade, they are at a passing level. So, I can give gentle encouragement to that student, telling them to keep working hard and that they are not doing terrible; College Algebra is a difficult class for many students at technical colleges!

Student D did have a derogatory remark, but I try to overlook that because they are being honest. This student had turned in no work at all due to financial aid issues. There could also be something going on outside of the classroom that I am not aware of. In any case, I responded to this student by giving her information regarding counseling services and encouraged her to talk to me or counseling services. That way, she knows she has someone in her corner wanting her to succeed. Yes, she may fail my class, but giving her resources, the college has a better chance of retaining her as a student, and she has a good chance of learning from her mistakes and getting that covenant degree so she can pursue her dreams.

Student E, on the other hand, is fighting hard to stay in college. He has told me in the past that this is his third time trying out college; in the previous attempts, he would just stop showing up and drop out. This gives me a great chance to encourage him further, tell him to keep up the good work. He disclosed he is working three jobs on top of taking my class. I wrote back to him that I realize that working three jobs and going to school is tough, but so far, he is proving that he _can_ do it. I underlined and italicized the word can on purpose. I made it an emphasis to uplift him, to encourage him to keep going, and to keep working hard, because I know he is fighting through a lot, but with more guidance, he will succeed and graduate to pursue his dreams.

Not only can exit tickets provide a great way for students to provide input on the class, but, as Wakeford suggested, it can be a great assessment tool as well. During my time as a graduate teacher assistant at Utah State, I would put a question on exams similar to this:

> Create an example and solve the example of a question pertaining to this exam that was not asked about already.

The idea is simple: students spend hours studying, and as an instructor, it is up to me to put on the exam the most important topics. However, that list may slightly differ from that of the students. So, simply putting a question on the exam asking them to solve a self-made example is a great way to have them show the instructor that they know more than just what was tested over. This idea is like an exit ticket, as an instructor I could gauge the answers to that question to see what the students felt was most important. That is valuable information, especially if the topic they think is important should not be as important. Maybe I spent more time on the topic then I needed to. Maybe students just love a certain topic over another. Maybe students understand different topics than the ones that were tested. In any case, questions such as the one mentioned provide valuable feedback to the instructor on how the course is being taught and what is of importance to the students.

6. Nurture a growth mindset mentality in and out of the classroom for your students.

With being a new parent, it seems as if everyone wants to tell me how to parent. At the time of this writing, I have a 4-month-old at home. While many people have told me the tips and tricks to parenting, it really all boils down to actually doing it yourself. I think teaching is a lot like this. No matter how many years you have trained to be a teacher or trained to be a mathematician, it takes time and experience to be the best.

As a parent, I want what is best for my child. When my child makes a mistake, I want her to be able to learn from that mistake. I want to help her develop a positive growth mindset in herself, that no matter what mistakes she does, Dad will always be there for her to walk her through what could have been done differently and how she can positively grow from her experiences.

As a teacher, I want what is best for my students. When students make mistakes, I want them to be able to learn from those mistakes. I want them to grow and develop into students who have a positive growth mindset and know that no matter what mistakes they make, I will always be there for them to uplift them and to show them that everyone in life makes mistakes. To make mistakes means to grow, and positive growth is, in my opinion, one of the best attributes anyone can have. So, teaching your students that failing is okay is critical to them succeeding in your class. Give them opportunities to grow from their mistakes, because even when they get a job, mistakes do happen, and a lot of times as long as the employee recognizes the mistakes and learns how to grow from it, they will be able to keep their jobs.

7. Plan activities that show students you want them to be successful.

Most students want to be successful, no matter what class they are taking. So, when they fail an assignment or quiz due to a multitude of different reasons, many of them feel discouraged and feel the need to cry, get angry, or any other range of emotions.

Having a positive mindset going into a class or an exam will do wonders in getting students to realize they can be successful in a mathematics course. We as instructors can help students be confident in their work by choosing activities to do in the classroom that are fun, engaging, and builds confidence. Here are a few ideas I have done in the classroom with some success.

After a cumulative midterm taken in class, the averages for two of my classes were 63% and 65%, which for a cumulative exam is not too bad, but I knew students would become discouraged when I passed back the exam. There were two activities I did for this exam before I passed them out. One, I had everyone tear out a half-piece of paper. The instructions for the activity were to write down why the student was attending CGTC. Was it because you are trying to provide a better life for your family? Was the student looking for a job promotion at work? Maybe the student is trying to prove their family wrong and that they can graduate with a college degree. I told them that no one will see what they write, not even me. It is completely up to them what they write, but I told them to keep this piece of paper with them throughout their time at CGTC. No matter if they fail an exam or are struggling with trying to juggle homework, jobs, and children, they can always look at this piece of paper to remind them why they are here. It provides them with motivation to keep going, keep trying, and not give up.

As they finished up, I even got a little emotional by telling them why I got a degree: I was a first-generation college student who loves mathematics and who loves teaching them mathematics. I told them I love my job because I get to help them be successful at CGTC and get them to wherever they want to do in life. Whatever they wrote down on that piece of paper, I want to help them achieve their life goals. It is my belief that by doing this activity, students realized that yes, they might have failed the midterm exam, but there is always a bigger picture at play. Keep your eye on the prize, because there are many more opportunities not only in this class but throughout their time at CGTC as well.

I also offered students the chance to correct their exams by redoing missed problems for half credit. So, not only did I give them a chance to reflect on their goals, but then I followed it up with a chance for them to develop a growth mindset. Activities such as this are a good way to show students that mistakes are okay; we all make them whether it is in a math class or in your career. Heck, I make mistakes all the time in my job. If I had a dime for the number of times I made a mistake while working a problem with my students in class, I could probably retire. But, alongside a point made earlier about working examples with a purpose, if we show our students that we can grow from our mistakes, then students will be motivated to grow and learn in a technical college math course.

8. Teach every day with a purpose and motivation.

Maybe this is another *duh* moment in this chapter. But, when you have been teaching the same class for 5 years in a row, how easy is it to lose motivation to teach? When students each and every semester make the same mistakes over and over again, at some point, do you as the instructor burnout? In 2017, the Learning Poly Institute did a study titled "Teacher Turnover: Why It Matters and What We Can Do About It." Just within the first couple pages can you find data supporting teacher burnout, especially in STEM-related fields. According to the report, "Turnover rates vary across subject areas," but it goes on to say, "Mathematics, science, and special education teachers have higher turnover rates, exceeding 13% annually." The report also goes on to say that "mathematics and science teachers often tend to have less teacher preparation than teachers of other subjects—in part because many enter through alternative pathways."[2] While this report is for teachers in primary schools, the same sentiment can be said for technical college instructors. I discuss alternative pathways into math education in technical colleges in the Appendix, so I leave this discussion for the Appendix.

With burnout being at an all-time high, being motivated to teach each and every day can be difficult. I think it is important to remember as an instructor why we do what we do. Our whole job is student-centered. Technical education is the backbone of America. Without us, students would not have an affordable option to try and get a college degree that will help them land a job postgraduation. Without us, many jobs throughout the country would be lacking employees. Without us, high school graduates would not have college credit after graduating high school. There is a multitude of different reasons to be proud to be a mathematics instructor at a technical college. Remind yourself each and every day of those reasons to be proud, and that will provide motivation and excitement to be in front of the classroom every day, teaching students not only about mathematics but also about how to be career-ready adults.

It is through my experience that each of the eight points I have made shows students that us as instructors want our students to be successful in and out of the classroom. We want to be that personal cheerleader for them. There is nothing that brings me more joy (well, besides my family) than grading an exam, knowing that the students have struggled, and seeing they received a passing grade. There is nothing that brings me more joy than seeing a student's face light up after "getting something."

Let me speak to the students specifically. We as faculty want you to succeed. I am here cheering you on from the corner each and every day. I look at your homework, I look at your exams, and it breaks my heart when you just are not "getting it." Please, take the time to come talk to me about a question you have about your homework. Please, stop me in class if you have a question. No questions are dumb, as they say. In fact, more often than not, the question you have someone else in the class does as well. Please, try your hardest because I know math is a difficult subject. Trust me, I have spent years studying it, and I hardly feel like I know anything. Finally, please know that when you finally get done with my class and you look back on all your successes and failures, that I hope you realize that no matter what you put your mind to, you *can* do it. You have the power to do whatever you set your mind to. If no one else, I believe in you.

So, student, go out there and change the world. All your instructors and I want nothing but the best for you. Show the world why you decided to take the plunge and get a college degree. Whether it be because you wanted to make a better life for your family. Maybe you wanted to prove to our family you are capable of such great feats. Maybe you are a first-generation college student showing your children what it means to be brave. Whatever the case may be, I want you to know that we as instructors will always be here for you. And know, no matter what obstacles you will face, you will always have one cheerleader in your corner.

REFERENCES

1. Wakeford, L. *Sample Exit Tickets*. The Harriet W. Sheridan Center for Teaching and Learning. Brown University. www.brown.edu/sheridan/teaching-learning-resources/teaching-resources/course-design/classroom-assessment/entrance-and-exit/sample
2. Carver-Thomas, D. and Darling-Hammond, L. (2017). *Teacher Turnover: Why It Matters and What We Can Do About It*. Learning Policy Institute.

Appendix: Alternative Pathways to Teaching Mathematics

I N Chapter 1 of this book, we discussed early on how individuals got their start in technical education. The majority of mathematics instructors have a degree in mathematics; that is what qualifies them to teach mathematics courses based on the standards set forth by the accrediting agency of the college. However, not all instructors begin in mathematics and do not have a pure math degree. This Appendix discusses some tips for those of you who are teaching mathematics at a college who do not have a pure math background. I also want to talk about the dynamics of teaching a mathematics class without necessarily having a teaching background (i.e., not having a math education degree).

Let me start with the second one about not necessarily needing a teaching background to teach mathematics at a technical college. In college, my wife always got after me, asking, "Why do professors get to teach college when they have taken no education courses!?" Honestly, she did not have a bad point. College instructors, especially professors at 4-year universities, are not required to have degrees that require education courses. So they did not learn how to create lesson plans, evaluate students in an efficient manner, or have student-teaching opportunities to help get them accustomed to teaching in the classroom. Well, that may not be fully true if a professor such as myself got teaching experience through their graduate program. However, if all a professor did was focus on research and not on teaching, then the last point still holds true.

At a technical college, research is not king. Teaching is. Let's take a look at the requirements on what it takes to become a mathematics instructor at Central Georgia Technical College (CGTC), taken from a job description post from when the college was hiring for a math instructor in May 2021.

To meet the standard academic qualification, an instructor must have earned a master's degree in the teaching discipline from a regionally accredited college or university or earned a master's degree with a concentration in the teaching discipline with a minimum of 18 graduate semester hours. Now, the post goes on to say this: education courses related to the teaching of Math are not considered discipline-specific in meeting this requirement. Coursework must be in Math or closely related discipline aligned to course competencies.

Interesting! So even if I took a math education course and I did not have a degree specifically in mathematics, I could not count that math education course toward the 18 graduate semester hours needed for qualification. Now, personally I did take a few education courses while doing my undergraduate. I even got a minor in education studies, which, according to the University of Northern Iowa website, is a program through which one can "explore the intersection of learning, education and society." It goes on to say that the minor will "increase your understanding of educational policy and get introduced to educational issues you may experience as a student, citizen, parent, or in your future careers."[1]

So what kind of classes did I take in this major? I took a class on the dynamics of human development, which speaks on theories and different viewpoints of human development from years 0 through 18. I took a class on the study of disability, which comes in handy when I have a student with a disability in my course. I took a class that explored the different kinds of higher education there are and how they play a role in American society. Outside of this minor, I also want to highlight a course I took named Technology of Secondary Mathematics Teachers, and while the name suggests the course being for high school instructors, I think it served me well as a college instructor learning about programs such as Desmos and GeoGebra.

Now, I don't want my listing off education classes I took in my undergraduate career to come off as "hey, look at me and how cool I am in taking all these education courses!" My point is that while instructors at the college level may not have education backgrounds, I would highly recommend those who are thinking about getting into mathematics teaching at any higher education institute to take an education class or two, or if you have been in the field for a while, seek out an opportunity to learn about some of these topics. For example, we all work with students with disabilities; why not learn a thing or two about how to provide a quality and equal education to these students?

I do think it is more important for instructors to get a mathematics background when teaching at a higher education institute than it is to get an education background. With an education background, as noted before, there would be lesson planning and teaching how to manage a classroom. The managing a classroom part, sure, that might be important (although many professors already assume students know how to act in a classroom). Lesson planning, though? Raise your hand if you are a math instructor at a higher education institute and have ever written out an entire lesson plan before. Mine isn't raised.

Now, this doesn't mean I don't come to class prepared. Of course I do. I look over the standards we need to meet in class, see what homework MyMath Lab has students working on, and tailor my lesson to meet the standards and what is on the homework. I have typed notes in LaTeX, and while I use them less often the more I teach a certain class, I still like to have them specifically to have preprepared examples.

For example, if I am teaching about factoring polynomials, it is good to have preprepared examples since not all polynomials can be factored. I need to be prepared so that I don't spend time trying to think about a polynomial that is factorable. Now, it may not be a terrible idea to teach students *how* to come up with factorable polynomials (start with an already factored polynomial then FOIL), but we would not want to do that for each example.

Going off topic slightly for a minute, I think it is a good idea to come to class with preprepared examples that are not already solved. As stated earlier in the book, it is important to be able to make mistakes in class. This shows students that everyone can make a mistake, even the instructor. With examples that are not worked out, not only can the instructor make a mistake, but the students can start to see how the instructor might tackle a problem as well. Even better, instructors can start asking students to make predictions about how certain problems might be solved.

Take, for example, a trigonometry class learning about solving trigonometric identities. Let's just pretend for the sake of the example we are trying to verify the identity

$$\frac{\sin(x)}{1+\tan(x)} = \frac{\cos(x)}{1+\cot(x)}$$

If I the instructor already have the solution worked out, I will be tempted to just work the example on my own. However, without knowing how the identity is verified, we can start to ask the students their way of thinking about the problem. Verification problems start by choosing a side to work on, then trying to make it look like the other side of the equation. So, the first question I asked my students: which side do they want to start on. Most said left-hand side. Great! Start with the left-hand side of the equation. "Now what," I asked my students. This is a great way to start assessing our students' knowledge. If they say cross-multiply and divide, we know right then and there that students are not understanding what it means to *verify* an identity. Minus this issue, if we start on the left-hand side, we hope students would say to use the identity

$$\tan(x) = \frac{\sin(x)}{\cos(x)}.$$

This would turn the left-hand side to

$$\frac{\sin(x)}{1 + \frac{\sin(x)}{\cos(x)}}.$$

Here is another assessing step: can students add fractions with unlike denominators? We can write the denominator with one fraction by using the least common denominator of $\cos(x)$.

$$\frac{\sin(x)}{\frac{\cos(x) + \sin(x)}{\cos(x)}},$$

which then simplifies to

$$\frac{\sin(x)\cos(x)}{\cos(x) + \sin(x)}.$$

To make a long story short, you would then need to perform a similar operation on the right-hand side of the equation so that both sides look the same. Through guided feedback, I was able to assess students' knowledge of the section and then have them get the answers themselves, which also boosts their confidence levels.

Speaking of assessing, education programs do have classes on proper ways to assess students. As a mathematics instructor at a higher education institute, I can personally tell you I have never once been taught how to make a "good" test, and honestly speaking, I'm not sure if any instructor has or not. Of course, there are many books and conferences on the topic, so I don't want to take much time in this Appendix to talk about them, but I do want to mention it is a good idea to be proactive as an instructor in the different ways to assess students, not only in the standard paper-and-pencil exam. Read up on them some time, and I can promise you not only will it spice things up in the classroom, but students will perform better in your class as well. One such example might be competency-based exams.

As stated earlier, taking the majority of mathematics courses instead of education courses is beneficial to mathematics instructors. I mentioned earlier in Chapter 5 about instructors needing to take real analysis before teaching a Calculus I course. One evidence that this should happen can be seen by viewing the list of topics normally taught in a real analysis course. These topics include

- sequences and series,

- limits,

- derivatives,

- integration,

and so much more. I wanted to list these four because what class are these topics introduced in? You guessed it, Calculus! Students in an introduction to real analysis course study the different theorems that are shown in calculus and see why they are true. Don't get me wrong, it is a difficult class (trust me, I have had four semesters of an Introduction to Analysis course, one semester of a graduate-level Real Analysis course, and had to pass a qualifying exam in my master's degree in analysis), but if we are standing up in front of a class teaching about topics that we ourselves do not fully understand, how do we expect our students to learn the material as well? I suppose, yes, one could teach limits without knowing what the epsilon-delta definition of the limit is, but to get a full understanding of the limit, one needs to see this definition.

In actuality, some calculus books even talk about the epsilon-delta definition of the limit. I had my Calculus I students perform an epsilon-delta limit problem on homework at Utah State University, and although I am sure they truly didn't grasp the full intention behind it, when they go onto higher-level mathematics courses, they will hopefully remember learning a little bit about it in Calculus I to aid them in that course.

Now, you might be saying, "Didn't you say that most students at a technical college do not take further math courses, so why does it matter?" And that's a good point. The fact of the matter is that even instructors at a technical college need to be proficient in the subject areas they are teaching, and if you are teaching calculus, taking an analysis course will help in accomplishing the goal of being proficient.

One other overlooked reason for why one should take these higher courses that can be translated to teaching at a technical college is the idea that you can learn some "fun" mathematics theorems that will produce curiosity. For example, in an abstract algebra or number theory class, you would learn about modular arithmetic. A fun way to keep a College Algebra class engaged from that? Ask them what 11 + 2 is. If they say 13, they are wrong! It's actually one! How? Look at the clock! Eleven o'clock plus 2 o'clock is 1 o'clock. So, 11 + 2 is 1! They get a nice laugh out of it, I can mention a fun math concept, and it keeps them engaged throughout the rest of the hour of class.

Another fun topic I tend to use in class is the idea of multiple levels of infinity. That one usually gets them to think hard! Without getting into too many specifics, mathematics has countable numbers and uncountable numbers. The set of integers is countable. The set of real numbers is uncountable. This suggests that the set of real numbers is larger than the set of integers, although both are infinite lists. So, one infinity is larger than another! The YouTube channel Numberphile has a wonderful video on this and a little bit of history on Georg Cantor, which I include the link in the references since I think it is a fun watch.[2]

One last topic I have fun with my students with is the idea of the wobbly table. Imagine you are at a table with four legs, and one of the legs is off the ground to make the table wobble. The mathematician way to fix this is to rotate the table until the legs become steady on the ground. This is an application of a corollary of the Intermediate Value Theorem (IVT)—if you have a continuous line where a is negative and b is positive, then the line has to cross zero somewhere. Yes, yes, the IVT can be more formally stated, but this summary should be a good one. This is a fun one to try at home, and I have in fact tried it at the local Target once, and it worked! Numberphile also made a video regarding this concept, which I have again included the link in the references if you are curious on watching it.[3]

Let's switch gears to instructors who do not have a mathematical background teaching mathematics at a higher education institute. Many of these instructors get their start in physics, engineering, or other science-related fields that uses mathematics. We could also clump in those who taught at a secondary school and then went on to teach math in higher education. This would be people who received some graduate credit in mathematics but not the 18 that is required to be credentialed.

First off, if you are thinking about becoming a mathematics instructor with not enough graduate-level credits to be qualified, you may be only able to apply to math instructor jobs that are for developmental mathematics classes. Sometimes called learning support, these classes are noncredit classes that students take to further develop their math skills before jumping into classes such as College Algebra. At CGTC, we have two active learning support classes, MATH 98: Elementary Algebra and MATH 99: Intermediate Algebra.

Just to give you an idea of topics and a description of each class: MATH 98 emphasizes basic algebra skills. Topics in the class include real numbers and algebraic expressions, solving linear equations, graphs of linear equations, polynomial operations, and polynomial factoring. MATH 99 is slightly different in the fact that it is a corequisite with our College Algebra class, meaning if a student is taking 99, they have to take College Algebra during the same semester. MATH 99 is a great way for students to get more exposure to topics that are also in College Algebra because the list of topics includes factoring, inequalities, rational expressions and equations, linear graphs, slope and applications, systems of equations, radical expressions, and quadratic equations.

CGTC also at one point had MATH 90. This class, titled Learning Support Mathematics, is a class that has not been offered since the spring of 2019 due to the lack of support for learning support classes and low enrollment. While it's not offered it anymore, I do want to mention it for those who are thinking about teaching learning support classes at a different college. This course has students work with whole numbers, fractions, decimals, percentages, ratio/proportions, measurement, geometry, and application problems. It also goes over an introduction to real numbers.

If these classes sound interesting to you, many colleges and universities have learning support classes in place to help those students who need that extra support. However, it may be hard to find a job! As of February 2022, searching HigherEdJobs.com for a mathematics instructor position with the keywords "Learning Support," I saw three job openings around the country. Not very many to choose from I must say! We currently have two instructors on staff who teach only developmental math courses, so it goes to show how small the world! However, in my opinion, it can be a great world to get into since one can switch the perspectives students have about mathematics.

Now, if you are someone who does have the qualifications to teach mathematics although your undergraduate degree may not be in mathematics, let me tell you a story about how great it is becoming a mathematics instructor. Let me start off by saying the pay isn't great. According to Indeed. com, the average salary an instructor at a community college makes in the United States is $54,542 per year.[4] So maybe not the best start to the story. However, I think any teacher or instructor in the nation do not necessarily go into the career for the money. Instead, it all starts with the students.

Yes, that's right. The students. Making $50,000 per year might not be enough, and it should be increased for every instructor out there, but I am still willing to do my job because students are worth my time and energy. They are the ones who are taking classes and putting in the hard work to get a college degree that they can then get a job in their dream field. When you get to see them walk across the stage at graduation (all faculty are required to attend graduation at CGTC), it's a great feeling to know that you, the instructor, were an integral part of that moment.

One of the best days and one of the worst days in the life of an instructor at a technical college is grade submission day. On this day, I get to see the successes and failures of each and every one of my students. Of course, I want every student to succeed; I don't think I should be in this job field if I didn't want that. However, there are always a few students you see pass that puts a big smile on your face. I had a student who was a nontraditional student pass my course, and at the end, after she turned in her final exam, took a selfie picture of her, me, and another student in the course. That other student in the picture came to me after the semester was over and gave me a Christmas card thanking me for a wonderful semester. These are the moments I live for; students are sharing with me their triumphs, and I got to be the piece that helped them succeed not only in my course but also in pushing toward their ultimate goal of graduating.

Maybe I will hurt myself again by writing this next paragraph in convincing you, the person who wants to teach math at a technical college, but as I said previously, grade submission day is also the worst day of the semester. I hate failing students. I really do. It's not something I look forward to because even if a student showed no progress through the semester, there is usually always something behind the scenes that caused the unsuccessful attempt at the course. However, even though it is something I don't like, it is something I have to do per college policy. So, with all the successes, you will always have the failures as well.

The rigor of mathematics can also be a difficult one to overcome for any new teacher of mathematics. It is a giant balancing act between keeping students engaged, but also teaching mathematics in a logical and cohesive manner. I have already talked about the rigor of mathematics in Chapter 5, so I won't go into too much detail again, but I do want to expand upon the idea of *how* to teach mathematics to a group of students at a technical college.

First off, there isn't one right way. You do you when it comes to teaching. You can't enjoy your job if there isn't freedom in how you perform your job. Some instructors board write, some use a document camera, and some use PowerPoint. Whichever way you teach, that is great! I haven't talked much in this book about classroom setups, but there is a big movement going on about traditional lecturing versus other teaching methods such as inquiry-based learning and flipped classrooms. The classroom will not be successful if the instructor is not comfortable, so no matter how or what type of classroom you run, make it your own!

When teaching mathematics, my biggest tip is to make sure your class has a story behind the class. I love to teach as if I am telling a story. For example, the shortened College Algebra story is this: we as a world want to describe everyday situations that happen to us. One way to do that is through the use of mathematics. We can write down equations to model situations in the real world, for example, profit on a business or finding the tax on a purchase. So, in the class, we learn about a lot of different equations that help us model these situations: linear equations, quadratic equations and polynomials, rational equations, radical equations, and exponential and logarithmic equations.

If you can sell the story and capture the students' attention, I believe you can have a very successful semester. Just like being caught up in a wonderful book, not every chapter or section you are going to like or understand. And that's okay. You keep pushing forward, and at the end of the story, you have successfully completed the book or the class and have hopefully learned a thing or two from it. That doesn't mean you are going to like the story when you walk out of the room for the final time, but the main takeaway is you were successful in the journey, and you will be rewarded with hopefully a college degree and a job afterward.

Also, one other piece of advice: listen to your students' and your employer's suggestions on how to improve. Especially if you are coming from a non-math background, it is important to take every professional development opportunity seriously, and one part of professional development is positive criticism. Whether that be through student surveys at the end of the semester or classroom observations by your boss, the main goal is to get better at what we do. We should be constantly striving to get better at what we do, and to do that, we need to listen. I always tell my students to take the student survey results seriously and to leave good feedback. Otherwise, how can I improve myself and the course for a future generation of students? Comments such as "you're great!" or "you suck!" (yes, I have gotten that comment before!) aren't productive criticism. Instead, comments such as "you're great because . . ." or "you suck because . . ." are a lot better because there is a reason behind those comments.

Like I said before, I am not the perfect teacher or boss, nor will I ever be. My point in writing this entire book isn't necessarily to tell you what to do or how to go about your day. The point was to give you a glimpse into the world of the technical college and teaching mathematics at one. There are many variables that are at play to create an environment in which students can not only be successful but also enjoy their time in both their mathematics courses and at the college.

So, if you do not have a mathematics background and want to teach mathematics, go and take a few classes to earn enough graduate credit to be qualified to teach at the higher education level! If you already have a math degree, consider going into technical college teaching. Not only will you gain a lot of experience in teaching mathematics content, you will get to meet some amazing people along the way who will help you go far in your career and in your life. As stated before, technical education is the backbone of the American workforce, and knowing that you are making an impact in these students' lives is life-changing.

REFERENCES

1. *Educational Studies Minor.* University of Northern Iowa. https://coe.uni. edu/epfls/majors-minors-certificates/educational-studies-minor
2. Haran, B. [Numberphile]. (2012, July 6). *Infinity is Bigger Than You Think—Numberphile* [Video]. YouTube. www.youtube.com/watch?v=elvOZm0d4H0
3. Haran, B. [Numberphile]. (2014, Aug. 18). *Fix a Wobbly Table (with Math)* [Video]. YouTube. www.youtube.com/watch?v=OuF-WB7mD6k
4. *How Much Do Community College Professors Make?* Indeed. www.indeed.com/ career-advice/pay-salary/how-much-do-community-college-professors-make

Index

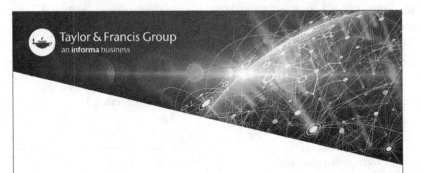

Printed in the United States
by Baker & Taylor Publisher Services